Backyard Chickens
BEYOND THE BASICS

PAM FREEMAN

VOYAGEUR
PRESS

Brimming with creative inspiration, how-to projects, and useful information to enrich your everyday life, Quarto Knows is a favorite destination for those pursuing their interests and passions. Visit our site and dig deeper with our books into your area of interest: Quarto Creates, Quarto Cooks, Quarto Homes, Quarto Lives, Quarto Drives, Quarto Explores, Quarto Gifts, or Quarto Kids.

First published in 2017 by Voyageur Press, an imprint of The Quarto Group, 401 Second Avenue North, Suite 310, Minneapolis, MN 55401 USA. T (612) 344-8100 F (612) 344-8692

www.QuartoKnows.com

Voyageur Press titles are also available at discount for retail, wholesale, promotional, and bulk purchase. For details, contact the Special Sales Manager by email at specialsales@quarto.com or by mail at The Quarto Group, Attn: Special Sales Manager, 401 Second Avenue North, Suite 310, Minneapolis, MN 55401 USA.

10 9 8 7 6 5 4 3 2

ISBN: 978-0-7603-5200-7

Cataloging-in-Publication Data on file with Library of Congress

Acquiring Editor: Thom O'Hearn
Project Manager: Jordan Wiklund
Art Director: Brad Springer
Cover Designer: Amy Sly
Interior Designer: Bradford Foltz
Layout: Diana Boger

Printed in China

To Allen, Lauren, and Kelsey:
you are my inspiration and my companions
in my chicken-keeping journey.
I love you!

CONTENTS

INTRODUCTION

I grew up hearing many stories from family members who raised chickens, but I never gave them much thought until I moved to the country with my own family. My husband had grown up in a household that kept chickens, and he thought it would be good for our kids if we added chickens to our new backyard. I did a lot of research, and then one day the Easter Bunny came along and surprised us with our first flock! Those chicks were a big hit. They grew to be happy and healthy—and we all loved the rich, flavorful taste of their eggs. Our backyard has been full of chickens ever since.

My chicken-keeping journey soon brought my career full circle, as I decided to combine my degree in journalism with my passion for chicks. I started a blog documenting the goings-on in my backyard and then began writing for the website of *Backyard Poultry* magazine. I eventually moved into a position where I fielded Ask the Expert questions.

The inspiration for *Backyard Chickens: Beyond the Basics* came from those questions. Although each query was a bit different, I noticed common themes running through all of them. Because my position requires constant research, I started to see the need for a book like this. There were plenty of beginner books that show people how to get started, and there were a few higher-level books that talk extensively of health, but there was nothing in the middle. That's where so many of the Ask the Expert questions had roots, and that's the content that this book aims to cover.

In this book, you'll learn:
- How to expand your flock and best integrate new birds
- All about the different colored eggs that grace your egg-collecting basket
- How to properly feed your flock as well as some natural chicken keeping tips to keep them healthy
- The inside scoop on common predators and predator deterrents
- How to tell if your chicks are hens or roosters
- How roosters affect flock dynamics

- About changing up your chicken keeping with the seasons so your birds stay safe and healthy
- And so much more!

You'll also notice a couple special sidebar features as you read. I tried to bring in many of the most popular questions I've received over the years. I also brought a few vintage chicken pieces for show and tell. From the first egg cartons to farm signs and salesman's pieces, I wanted to share a little of the rich history of chicken keeping in this country. After all, everything from zoning laws to world wars and even good old fashioned advertising influenced backyard chicken keepers.

I think there's something for every chicken keeper in *Backyard Chickens Beyond the Basics*, from the beginner to seasoned keeper. I hope you (and your flock) enjoy the book and learn something new!

—Pam Freeman

CHAPTER 1

Expanding
Your Flock

Adding new flock members is irresistible. Each year, spring starts to come around and we chicken keepers find ourselves saying things like, "A few more chickens won't make much difference," or "I'd really like to try this new breed!" Before you know it, that first flock of chickens has grown, and you're thinking about adding an extra coop to your backyard set-up. (This is a phenomenon known as *chicken math* that often happens to chicken keepers, even the most seasoned among us!)

I should know! For me, the chicken-keeping journey started when the Easter Bunny brought four beautiful Silver Laced Wyandottes. Four was a reasonable starter number to be sure. Beginners at the time, my family found these chickens allowed us to try out all of our new book knowledge and dive into chicken keeping without being overwhelmed.

Yet there were others in my extended family who had raised chickens before, and they said we needed a few more chicks. They told us chickens aren't that hardy and that a small flock can die in one fell swoop! Instead of going through these proclamations calmly and rationally, I was led to the local feed store, as if adding to my chicken flock was going to inoculate me from bad things happening. (And really, I thought, who doesn't need more eggs? I could sell them!)

So there I was in a feed store, with my kids, adding new chicks to our growing feathered family. I had read up on my breeds, and my husband told me to go ahead and buy anything that looked interesting. Let me tell you, that's a bad piece of advice. Those chicks were all so cute and interesting! I guess my math skills escaped me and my rational thought flew to another planet: by the time I left that shop, our easily managed flock of four Wyandottes grew to a flock of 19, including New Hampshires, White Leghorns, Partridge Cochins, Easter Eggers, and Barred Plymouth Rocks.

This is how I learned chicken math the hard way. While all our chicks lived and we managed to get by, I quickly found that impulse buying was not the way you should add to a flock.

Nowadays, I don't go shopping for new chicks every season. If it's a season where I don't need

Take stock of your existing flock before adding more chickens. Evaluate your needs and what you have on hand, like existing coop space and free-range areas.

new chicks, I stock up on all my supplies ahead of time. And I try to avoid any of the stores where I know they'll be selling chicks for about two months! However, even for a seasoned veteran with a plan, things happen. Invariably you'll find yourself online with credit card in hand, placing an order for just a few new additions. What's one or two more, after all? Or you'll find yourself at the feed store staring at a brooder full of adorable balls of fluff. (That's how we got our accidental rooster!)

Even for those of us who never truly learn, it's important to step back from time to time and evaluate your flock. Take a breath and examine the chicken math before making an impulse buy. Your family and your chicks will thank you for it later. In this chapter, we'll take a look at expanding flocks, starting with the important concept of chicken math.

Considerations Before Expanding

Let's start by thinking of the oft-repeated phrase, "What's one more chicken?" On one hand, people who say that are right. One more chicken is usually no big deal. Yet the reality is it's never just one more. Chickens are flock animals; they need friends. If you're buying chicks, most localities even have laws to prevent folks from buying just one chick. The laws vary from place to place, but you're usually required to buy a minimum of three to six chicks when you make a purchase. Likewise, if you're hatching eggs, you're not going to put just one egg in the incubator.

Cost

As you might have guessed from the preceding text, the first thing to consider before expanding your flock is whether you have the ability to handle the increased expenses—and the work of additional chickens. I know each time we expand, I soon notice that I'm making trips to the feed store a little more often and buying more when I get there. An extra bag of food, an extra couple bags of bedding, and a few more treats . . . it all adds up.

Here's an idea of what you can expect, though keep in mind this will vary depending on the breeds you're raising, the time of year, whether they free range, and whether they're laying eggs or being raised for meat. On average, each baby chick will go through 9 to 10 pounds of feed for the first 10 weeks of life. That's about a pound a week. Adult chickens eat about 1.5 pounds each per week. So if you add six baby chicks, you'll need at least two 50-pound bags of chick starter feed to cover the first two months. When they're eating layer feed, after 18 weeks of age, you'll need an extra 36 pounds of feed for your additional chickens per month. True, a bag of food may cost just $15. But consider buying one more bag of food per month for a year: that's an extra $180.

Likewise, if you purchase bagged wood shavings, those cost around $7 per bag. It's hard to estimate the extra use of shavings, because that depends on the size of your coop, how often your chickens are in the coop, how thick you lay your shavings, and how often they are cleaned. I use at least four 8-cubic-foot bags of pine shavings when I clean the coop. I also keep an extra bag or two on hand for emergencies. When I purchase

This Speckled Sussex chick is getting some much-needed rest after settling in to its new home in the brooder.

six new chicks, I figure on at least one extra bag per month. That's an extra $84 per year. Taken together, you're now spending a total of $264 more per year before treats and other costs. Increased work is a little harder to calculate, but more chickens will certainly take more time out of your schedule. You may need to add another waterer, for example. And each day that extra waterer has to be cleaned and refilled at least once. Or you'll notice the coop will get dirty a little more quickly. That means you'll be taking a little more time on small daily cleaning chores and performing a total coop cleaning a little more often.

Space

The second thing to consider is whether you have the coop space to handle more birds. This will vary from backyard to backyard and situation to situation. If your birds free range all day and only use the coop for nesting and egg laying, you need a much smaller coop than someone whose flock stays confined all day. No matter where you fall on that spectrum, it's always a good idea to build or purchase a bigger coop than you need at the moment. It's only natural that you'll want to add more birds at some point. It's much easier to have the space on hand compared to having to retrofit an existing coop or add another coop entirely. If your birds don't free range, more space is always better. So it's a win-win all around.

Age

The third thing to consider is the age of your flock. Standard chickens can live 8 to 15 years. Bantam chickens can live 4 to 8 years. As chickens get older, their egg laying decreases. It's a myth that hens just completely stop laying eggs all at once. Hens are at their most productive for their first two to three years. After that, their production just gradually tapers down. I find that for many backyard chicken owners, this isn't a huge problem as most of us have a flock with chickens of varying ages. That means at any one time most flocks have newer egg layers and older hens tapering off production. It evens out for fairly consistent production, so many chicken keepers aren't worried too much about who's laying and who's not.

However, there are situations in which this isn't the case. If you're raising chickens so you can sell eggs, then you don't want a flock of older hens. You'll want to plan out retirement options for the older ladies while timing it so there's not a drop in production as your new birds are growing to a producing age. For the backyard chicken owner, the age of your flock is probably most relevant in neighborhoods with tight regulations on chicken keeping. If you're constrained by regulations, you don't have the ability to just add to your flock. In this case, careful thought needs to be given to how your flock will be managed over the years.

There's no right or wrong answer to all these considerations. Each person's situation will be different. But they're important to keep in the back of your mind when you're thinking about adding to your flock. I mentioned that we once bought a rooster by accident through a feed store purchase. In that case, I didn't make an impulse buy. I knew I wanted to add to my flock. And as I stood there, with kids and husband in tow, I ran through all these thoughts in my head. Yes, I was prepared for more work. My coop and yard could handle more chickens. (After all, at one point we had 23 chickens happily living in it.) Adding new chickens wouldn't take my flock anywhere near that large number. And this purchase would give me a good mix of hens of varying ages. Or so I thought….

Hand-Raising Chicks

You've thought long and hard and you're ready to add to your flock. Great! Now we'll take a look at the two primary options when it comes to expanding an existing flock: hand-raising and letting a broody hatch eggs. We'll also consider buying older chickens. Each option has its own pros and cons, requirements, and required skill levels. But each option can be successful, rewarding, and fun!

Hand-raising day-old chicks is probably the most popular way to expand a backyard flock. I believe it's something every backyard chicken owner should know how to do. There are three main ways to acquire chicks to hand-raise: buying from a feed store, purchasing directly from hatcheries, and buying and incubating hatching eggs.

Hand-raising chicks is something every chicken owner should know how to do.

Buying From a Feed Store

Buying chicks from a feed store couldn't be easier. Each spring they're readily available in local stores across the country. I find it's best to get the chicks as early in their lifespan as possible. By doing this, you get more time to bond with them. I've found this results in adult backyard chickens that are much more likely to be friendly and ready to interact with humans. (This is especially important for families with children.) It's good to stay in contact with your local feed store in the spring so you know when they'll be getting their chick deliveries. That way you can swoop in and buy the chicks at the earliest possible time.

People often ask me what to look for in feed store chicks. I advise looking for chicks that are active and alert. I check to see that they're eating and drinking. I look for down feathers that are clean and well groomed. I look to see that their rumps are clean and free of poop. I also check their extremities: Are their legs straight? Can they walk well? Are their beaks straight and not scissored (crossed)? Do they have clear eyes?

A sick chick will be hunched over, or sometimes they'll stand still and sway with their eyes closed. However, be careful about this as healthy chicks need to sleep as well, and they like to get comfortable when they nap. So you may see a chick that's perfectly healthy but curled up, sleeping—or for that matter, all spread out dozing. Don't confuse natural sleeping with being sick! Here are a few additional tips from a feed store chick-buying veteran.

- If the feed store is putting your chicks in a box, have them line the bottom with some clean shavings. Normally there's an open bag nearby and the feed store is happy to add a handful to your box. This makes the slippery box surface a lot less slippery and messy—and it gives your chicks a comfortable ride home.

Healthy chicks are usually busy eating and drinking in feed store brooders.

- Although the bins at the feed store are marked pullets (girls) and straight run (boys and girls), this does not always hold true. First, hatcheries are not 100 percent accurate in their sexing. So there could be some roosters in there from the start. Also, a well-meaning employee or customer may set a chick back in the wrong brooder after looking at it. So you may end up with a surprise or two in a few months!

- Make sure to double-check your chicks once they're in your box. Most feed stores don't allow returns. So, if there's a problem with your birds, the time to address it is right then. Once you walk out of the store, you have to live with what you've got.

Purchasing Directly From a Hatchery

I'm lucky that where I live we have a well-known hatchery less than an hour away. I have the ability to place my order directly and then go pick up the chicks! That means there's no mailing and just a short car ride for the chicks instead. The other nice thing about having a hatchery close by is that most of our local feed stores get their chicks from this hatchery. This means I'm reasonably sure I'll get quality chicks when I purchase from my feed store, too!

There are some advantages to ordering directly from a hatchery even if you don't live quite as close. You can time your purchase to meet your

Q&A: Hand-Raising Chicks

Do chicks need to roost?

Chicks don't have to roost, but it's fun for them. Provide a small roost for the chicks in the brooder. You can get creative with the roost. I've used Legos to hold up a small stick horizontally and then added to the Legos to increase height. If you're using a dog cage, you can pop the stick through the holes. No matter how you create your perch, it will get your chicks used to the idea of perching and hopefully keep them from perching on inappropriate spots like the top of the water container.

What's the best type of brooder to use?

People get creative with their brooders. A large plastic storage container works great. Chicks grow fast so you'll need that extra space quickly, plus keep the lid and you can use that as your storage container for all your brooder supplies. For very tiny chicks, an aquarium works well until you can find something bigger. Puppy playpens work well for larger chicks. Stores also sell cardboard brooder set-ups.

What's a minimum number of chicks to start with?

You'll find that each locality and each hatchery has rules on the minimum number of chicks you can purchase. That's your starting point. If you have no minimums, it's best to start with at least three chicks. Since chickens are social animals, three chicks gives you a small cushion in case one doesn't make it.

When can the chicks have other food than commercial feed?

You can feed chicks treats in the brooder. Treats are just that, treats. So they should be fed sparingly. The bulk of a baby chick's diet should be a good commercial feed so they have enough nutrition for proper growth. Use common sense when feeding treats to chicks. Large pieces of food are hard for chicks to eat and digest. You can use the list of good chicken treats in Chapter 7 as a guide. If the food is larger, cut it up into small portions. Be careful with scratch grains. Some varieties come with big grains that are just too big for a small chick. The one rule for feeding treats to chicks is to make sure to give them access to chick-sized grit if they eat anything other than commercial feed. The grit will allow them to properly digest their treats.

What temperature can chicks go outside?

Around three weeks old, you can take your chicks outside on a nice warm day; about 70° Fahrenheit, sunny and not windy. Around six weeks old, the chicks should be feathered enough to move outside. These are guidelines, however, since some chicks will feather out faster than others. If they're moving into their outside coop, I advise waiting until the nights don't get below 50 to 55° Fahrenheit outside the coop. Inside should be warmer; especially with younger chicks still growing in feathers.

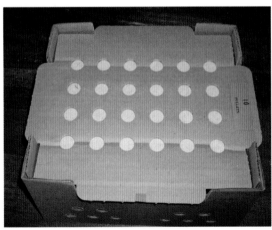

Hatchery chicks come in a fully insulated box that's designed to keep them safe during shipping. If you pick the chicks up directly from the hatchery, they'll be in the same type of box.

schedule. If you're not ready when the feed stores are, that's OK. Chances are your hatchery of choice is offering chicks almost year-round.

It's also more likely that they'll have a wider variety of breed options available at one time. For instance, if you want a couple Easter Eggers and Plymouth Barred Rocks plus a few bantams, your feed store may not have all those choices available at the same time. And, here's the rub, with local regulations about how many chicks you can purchase at one time you might have to go back to the store and purchase six chicks at a time over and over again to get the varieties you want. Or you'll have to settle for not having your wish list completely filled. With a hatchery, you can usually order multiple varieties at the same time. Ta da! No compromises on the wish list!

If you don't have a hatchery close by and you have to order through the mail, then the disadvantage is that you are often required to buy large amounts of chicks at one time. That minimum number can be as high as 25 chicks! Many backyard chicken owners don't have the kind of space to accommodate those numbers. But don't let this stop you. Many times feed stores will take special orders and process them when they're placing their bigger store orders. That way you're piggybacking on a larger order, and you don't have to meet the minimum requirement on your own. Your order will still be separated from the rest; you just have to pick your chicks up at the feed store rather than the post office. There are some hatcheries that will allow orders as small as three, though, so make sure to do your research. I've also found you can often find someone else in your area who wants chicks and you can combine orders to meet larger minimums.

Mail ordering has been a staple in this country for many years, and it's a great option. But it's good to remember there's a chance that not all the chicks will make it healthy—or alive— to your house. If you have kids, make sure you open the box by yourself first before unveiling the new chicks to your kids. That way you can sort through them first and then let your kids see them happy and settled in the brooder. This sometimes rough nature of mail order is one reason I like purchasing chicks at the local feed store. They've opened the box, sorted through the chicks, and vetted them. I remember one time I stopped in a feed store during chick season and the employee was crying over the brooders. The chicks in their shipment had all piled on top of one another crushing and smothering the ones underneath and in the middle of the pile. She was taking them out of the box and there were so many lifeless or dying bodies. I helped her sort through them and get the ones that had a chance of recovery to a safe comfortable place. It was heartbreaking. Just imagine if that was your own mail order box.

Incubating Hatching Eggs

Incubating hatching eggs is a wonderful choice for adding chicks to your flock. There are lots of breeds that aren't available as day-old chicks that are available in hatching eggs, so it's an especially good choice if you're looking to explore more breed options. It's also very fun for kids to see the chick developing in the egg and then hatching. Incubating, as you might expect, does require an incubator. Their prices and quality vary, so it's good to evaluate your options. Will you be using the incubator time and time again or do you plan to use it just this once? You may have a chicken-keeping friend who has a brooder and isn't incubating at the moment and will let you borrow it. Regardless of how you get the equipment, it's important to realize that this is the most involved way to get baby chicks. Depending on your brooder, you may be required to turn eggs each day. You'll need to be on hand through the incubation process (around 21 days) and then around to hand-raise the chicks. So no vacations for a while! And, you'll need to be able to handle any problems with the chicks, from hatching issues to deformities. Also, there is no guarantee of the sex of your chicks. Statistics show that more than 50 percent of chicks will be roosters.

Hand-Raising Considerations

Hand-raising chicks is fairly straightforward and pretty easy as long as you follow some simple rules and let your chicks be your guide. Here are some of the main considerations.

Brooder Size: It's important to realize chicks grow fast. Make sure you have enough space for them as they grow. The minimum square footage for a chick is 2 square feet.

Warmth: Heat is vital. A brooder should be kept at 90 to 95° Fahrenheit for the first two weeks and then moved down five degrees per week until the chicks are a month old. The temperature of the brooder should be taken at the level of the chicks so you know what they are feeling. Your chicks will also let you know if you've got it right by happily eating and drinking and chirping to themselves. If they're all huddled together and squawking, then they're too cold. To provide heat, you've got a couple options. You can use a standard heat lamp and place it about 20 inches or so above the brooder. You can raise or lower it

to get the temperature just right. If your chicks are all spread around the sides away from the light, then they're too hot.

> ### Note
> It's important to use fire safety rules with the heat lamp. Make sure it is well secured and kept a good distance from anything flammable. There is also a shelf heater produced by Brinsea called an EcoGlow. This comes in sizes according to the number of chicks you're brooding, uses less energy than a heat lamp, and is generally considered less of a fire hazard. Its cost is more initially than a standard heat lamp, but I use it and highly recommend it.

Food and Water: Chicks need food and water, but when you first get them they won't know how to drink and eat or where to find what they need. You'll have to show them how to eat and drink by dipping their beaks gently in the water and food. It's best to introduce water first. Make sure they're all getting the hang of drinking and let them get hydrated for about an hour. This also helps prevent pasty butt. After that, introduce the food and make sure they've all got the hang of eating. Your food and water containers should be chick-safe as well. There are a number of chick feeders and waterers available at your local feed store. (In a pinch, you can use a small dish with pebbles in it for water and an egg carton tipped upside down for food. Just be careful and keep a close eye on them.) As the chicks get older, you'll need to raise the containers so they're not constantly getting dirty, soiled with bedding and chicken droppings. You can get creative in raising the food and water containers. Landscape pavers work well and provide a sturdy base for the containers to sit securely. As your little ones grow, they'll definitely try their new wings and enjoy sitting on top of the containers. Be sure to check the containers often and be prepared to do a little extra cleaning since your birds won't be shy about where they leave their droppings.

Bedding: There are lots of options here, but I prefer using wood shavings just like I do in my adult coops. If you choose to do the same, add a generous layer and then remove soiled bedding as needed. A couple things to remember: Do not use the wood shavings that are fine like dust and do not use newspaper exclusively in your brooder. Chicks will eat the fine shavings and that can cause digestive problems. Newspaper on its own is too slippery and can cause splayed legs.

Using a Broody Hen to Add to Your Flock

A broody hen, simply put, is a hen that wants to hatch eggs and raise baby chicks. They are great to have around if you're looking for an easy way to add to your flock. A broody hen will gladly relieve you of that work!

Not every hen will be a broody hen. When a hen goes broody, she stops laying eggs. This is

A broody hen can be an easy and efficient way to add to your flock. Provide a safe place to incubate and hatch her chicks, and a broody hen will do the rest.

not considered a desirable trait in an industry that values maximum egg production. So if you want a broody hen, you have to choose your breeds wisely. I've put together a list that shows which common—and not so common—breeds will go broody. It's important to note that you will find discrepancies between breeders about which breeds go broody and which don't.

Breeds That Go Broody

As with humans, not all chicken personalities are the same. You may have a hen whose breed is well-known for broodiness, but never goes broody herself. The two breeds from this list that are probably the most famous for going broody, and being good mothers, are Cochins and Silkies. Both are easy-going, family-friendly breeds that make a great addition to a backyard flock.

SETTER

- Araucana
- Aseel
- Australorp
- Belgian Bearded d'Uccle
- Brahma
- Cochin
- Cubalaya
- Delaware
- Dorking
- Easter Egger
- Holland
- Houdan
- Java
- Jersey Giant
- Langshan
- Malay
- Marans
- Modern Game
- Naked Neck Turken
- Nankin (bantam)
- Old English Game
- Olive Egger
- Orpington
- Phoenix
- Plymouth Rock
- Shamo
- Silkie (bantam)
- Sumatra
- Sussex

VARIABLE

- Buckeye
- Chantecler
- Dominique
- Egyptian Fayoumi
- Faverolle
- New Hampshire
- Rhode Island Red
- Dark Cornish
- Wyandotte
- Welsummer

Signs Your Hen Is Broody

Broody hens are pretty obvious in a flock, and it's amazing how fast it can happen. We had a Partridge Cochin named Hoppy that was perpetually broody starting each spring and continuing through the fall. One spring my husband decided he really wanted our kids to experience a mama hen and her chicks. Hoppy complied by going broody and 21 days later Hoppy became a proud mama hen. With Hoppy, she'd be a completely normal chicken one day, and the next day she'd be a broody hen. In fact it happened so much even my kids knew the signs. They'd come back in after visiting the coop and say, "Mom, Hoppy's broody again!"

The first sign a hen's ready to set is when she grabs a nest box and doesn't leave it. If you walk up to the nest box, she'll start growling, sometimes hissing and making low guttural noises. If you try to remove her from the nest, your formerly sweet hen may peck you. Once out of the nest, she'll keep making low growls and puff all her feathers. It can be downright daunting to hear and see a broody hen outside her box! We were lucky with Hoppy; she was sweet and not too temperamental when she was broody. We could pick her up, and often did, and she'd grumble but never hurt us. Other hens become

If you want to use a broody hen to expand your flock, make sure you have breeds that will go broody. Orpingtons (above) are friendly hens that are known to go broody. Cochins (below) are one of the most reliable breeds to go broody on a regular basis.

monstrous. They'll bite you. They'll try to steal eggs from the other nest boxes, even when the other hens are in them. They are intent on doing their job and protecting their brood. Everyone else needs to stay out of the way!

Note

Don't worry if your broody hen has a bare breast, as a broody hen will pull feathers from her breast and use them to line her nest. This is much the same concept as a down coat in the winter. Those feathers will insulate the nest to increase humidity and supplement the broody hen's body temperature, which will rise so that she can fully incubate her eggs.

Once you see the signs, you need to decide what to do with your broody. For Hoppy, we left her in her nest box and moved everyone once the hatch was complete. If your broody has decided to nest in an unsafe or inconvenient place, then it's a good idea to move her to an out-of-the-way nest box or set her up in a broody pen that will be safe for both mom and chicks. Moving does have its drawbacks, though, because it can actually break a broody. Be sure to carefully move the eggs and try to set them in the nest the same way your broody had them in the first place. Sometimes broodies can get picky and will kick out eggs that seem different.

The Broody Hatching Process

It takes an average of 21 days for chicken eggs to hatch. I've found it's a good idea to mark each egg that the broody is hatching. That way you don't have any mix-ups with other eggs that end up in the nest. Use a thin line Sharpie permanent marker, so the markings won't disappear after a few days. Mark the date your broody starts sitting on her eggs in your calendar. That way you'll have a good idea of when to expect they'll hatch. It's important to understand that this is just a guideline. All your broody's eggs weren't laid in one day, so it's likely they won't all hatch in one day. Many chicken keepers have gotten to the 21st day with no hatch and gotten rid of all the eggs. That's a big mistake! It's actually similar to due dates when women are pregnant. The date is more of a guideline than a rule.

If you don't have a rooster but have a broody hen, you can still let her do the work by ordering hatching eggs and placing them beneath her. You can also buy day-old chicks and put them beneath a broody. However, use caution with doing either of these things. Make sure your broody is a determined broody. Some will sit on eggs and then wander off for a few days and then return again. Broodies need to be consistent with their setting. Also, make sure your hen is an accepting hen. Most broodies will willingly accept hatching eggs, but not all broodies will accept day-old chicks. It's often best to make introductions when the broody hen has been setting a while—and it's nice to have a comfortable area that's safe with low lighting. Once introductions are made, monitor the situation closely. You'll know if the broody is accepting the chicks when she lets them get under her wings for warmth. She'll also start tidbitting, which is a small clucking sound that tells the

A broody hen will do all the work of raising chicks, from teaching them how to eat and drink to keeping them warm and entertained.

chicks it's OK to eat and drink here. However, be ready to step in and assume the role of mom if your broody isn't doing the job.

BROODY TIPS

• Pictures on the Internet of broodies trying to sit on huge numbers of eggs are entertaining but don't necessarily represent a successful clutch. Hens can cover anywhere from 12 to 18 of their own size eggs.

- Place a small heat source on one side of the broody's cage. That way if the broody rejects her chicks during the night, they'll have somewhere to set up camp.

- A good rule of thumb is that a mama hen can raise as many as three times more chicks than she can hatch. So, if you have another broody who doesn't live up to the task, many times another broody will finish the job. Broody hens have even been known to foster birds of another kind, like ducks and turkeys.

- Make sure that you are ready to handle incubating eggs or hand-raising chicks. Sometimes things don't go as planned and a broody leaves her eggs in the middle of a hatch cycle. You can incubate those eggs with success. In fact, I've seen studies that show incubating eggs can go as long as 15 hours without heat. It may delay hatch, but you may still have success. Also, just because a broody successfully hatched eggs, doesn't mean she'll automatically be a good mother. Be ready to step in if needed.

A Broody Timeline

Days 1 through 17: The broody will turn and move her eggs for optimum temperature control and chick development. The broody will get up from the nest to eat, drink, dust bathe, and relieve herself each day. Broody hens have a special broody dropping that's instantly recognizable. It's huge and smelly—that's because the broody saves everything up and goes all at once!

Days 18 through 21: The broody will move her eggs into their final hatch position and she will not leave the nest. Try to not to pester the broody too much during this time. Don't move the eggs and don't clean the nest box. A mama hen will softly cluck during this time. She can actually hear her chicks and they can hear her. She offers them encouragement and comforts them as they're getting ready to hatch. If you listen closely, you can sometimes hear the chicks chirping too. (More on chick vocalizations on page 40.)

Beyond day 21: What about un-hatched eggs? Amazingly, the average successful hatch rate for a broody hen is almost 100 percent. But again, leave things to the broody hen. She knows when all the eggs that are going to hatch are hatched. She will usually kick eggs out that aren't going to hatch. Once she gets all the chicks up and out to find food and water, you can remove the un-hatched eggs. If all goes well, you'll have a broody hen who will hatch her eggs and raise her babies without much help from you. She'll teach the chicks how to drink, eat, and behave. She'll also make introductions to the flock when the time comes.

Adding Older Chickens

There are quite a few opportunities for adding older chickens to your flock. Some hatcheries will sell older chickens. Most people, however, end up acquiring older chickens by swapping with friends, through poultry shows, or by adopting rescue hens. There are a few tips to help you

Chickens are curious about newcomers and like the chance to get to know their new flock mates.

successfully add older chickens to your flock and make sure you're happy with the addition too.

Chickens Don't Show Their Age: Chickens aren't like humans. With humans, you can often make a reasonable guess at age. With chickens, not so much! Once those adult feathers are present, chickens look pretty much the same. You may ask why this is important. If your chickens will primarily be pets, then it's not. But, if you want consistent egg layers and you get a flock of older hens, you may not get the number of eggs you want or need.

Note

Introduce more than one adult chicken at a time. There's security and comfort in numbers!

Look Out for Diseases: Chickens from outside your flock are a bio-security nightmare. Even if the chickens you're getting don't *look* sick, they can be a danger to your chickens and vice versa. This is because each flock of chickens gets exposure to different diseases over time and develops

their own immunities and resistance. That exposure will vary from flock to flock depending on your geographical location, whether your flock free ranges or is primarily located in their coop and run, and other factors. So if your new chickens carry diseases but are not sick from them, and if your flock has not been exposed to the same diseases, then they are in danger. I think of this like getting a new job. Your co-workers all look healthy, but you can expect to get sick shortly after being around them for a while! It's the same thing with chickens, and it can lead to major heartbreak. I recommend quarantining your new birds from your old birds for at least 30 days. This will give you a good idea of the overall health of your new flock and let them start to get used to their new environmental factors as well.

Older Chickens May Need to Develop Social Skills: This happens primarily to hens that have never lived in social situations whether in a factory setting or a caged environment. They not only need to be quarantined for disease purposes, but also need time to learn how to live in their new surroundings. They will probably never have perched or used nest boxes, for example, so it's a good idea to make sure their quarantine area is set up like a normal coop. They will have the right instincts; they just have never been allowed to use them! These hens can be missing feathers from being pecked by others or from stress and self-pecking. Exposed skin can be a magnet for pecking and can easily be broken even when the pecking isn't aggressive. This means battery hens will need to re-grow feathers and get healthy before joining your main flock. It can take time and patience, but many people find it rewarding to see battery hens get a new lease on life.

CHAPTER 2

Flock Behavior

Until I started raising my own backyard flock of chickens, I never thought much about chicken behavior. Yet after I got my birds, I found myself entranced. It started with the chicks in my brooder. They were fascinating! I spent hours watching them scratching, pecking for food, grooming, and even learning to perch. Once they were grown, I loved going outside and interacting with them.

My Barred Plymouth Rocks were the best foragers. Any time I dug holes for planting or turned over a rock or log, they were there to get the best goodies. My White Leghorns, contrary to their breed profile, were so docile I could hold them in my hand and pump them up and down like I was weight lifting.

Hoppy, our Partridge Cochin, formed a special bond with us. She loved to talk to us and come over for pets. One day our dog, Sophie, got out of our fenced-in yard. We needed to get her back safely so we split up places to find her. I went to the bottom of our driveway to make sure Sophie didn't get out onto the road. My husband stayed in the backyard area calling for her and monitoring the front of the house too. I had been gone a while so I came up the hill to report my lack of progress. There I found my husband calling for Sophie and right next to his feet was Hoppy. Every time he would call, she'd call out too. (I definitely think Hoppy helped since Sophie returned to the backyard on her own!)

Now, years later and with much more experience, I confess I'm still fascinated by chicken behavior. I've found chickens are definitely smarter than people think. They are capable of learning basic routines and adjusting to meet their needs. Our New Hampshire named Big Red knows that we have food inside the house. She's well aware of where the door closest to the kitchen is located and will make her way there as much as possible. Once there, she'll call loudly until someone hears her and either shushes her away or lets her pop into the mud room to grab a treat. She learned this because each time she came to the door we gave her a treat. We thought the whole thing was fun, and we unknowingly established a routine. Red can also figure out where voices are inside the house and knows if she creates a ruckus then we'll come outside and check on her, usually resulting in a treat being given. I first noticed this when my husband was in our bedroom on the phone. Big Red was smart enough to walk around until she heard his voice and then stand below that window and call loudly. After he got off the phone, he went outside to see if she was OK and, sure enough, she got a treat. Smart girl!

Perhaps one of the best things about owning chickens is the countless hours of entertainment and fascination they can hold for the whole family. Let's explore the world of flock behavior!

The Pecking Order

Chickens are flock animals that enjoy social interactions. As with any group, they have a way of organizing so that order is maintained. This is called the pecking order and it influences the daily activities of the flock, from eating and drinking to perching and dust bathing. It has been theorized that the pecking order started

This Easter Egger (front) is raising her hackle feathers and puffing out her chest to let the Buff Brahma (back) know that she is in charge.

with Red Jungle Fowl in Thailand. When food was found, it was important that the flock stayed quiet and orderly so they did not attract the attention of predators. The highest-ranking birds eat first and then lower-ranking birds eat. That way the strongest birds remain fit and able to reproduce, passing on their strong genes. In a flock of chickens, the dominant bird is at the top and no other bird is allowed to peck that bird. However, the chicken at the top can peck all the others to tell them what to do. The pecking order descends like this from highest to lowest in

rank, with the lowest bird not being able to peck any of the other birds while all of the other birds are able to peck him or her.

The pecking order in a flock is established early. In fact, studies have shown that chicks can start to show competitive behavior at three days old. After they are 16 days old, they begin to establish the order of dominance. With an all-hen group, the pecking order will be set by the time the chickens are 10 weeks old. It can be even earlier for a small group of birds—possibly as early as eight weeks.

Note

Pecking is not always bad or violent. It is a normal and important form of communication. In fact, pecking is usually gentle and not even all that noticeable by humans. You'll find feathers are rarely disturbed as chickens "check out" each other and establish a hierarchy for functioning as a group.

Besides pecking, there are other ways chickens work out their order and show dominance. One chicken might challenge another by puffing up her chest, standing tall, and flapping her wings. The challenged bird can then either choose to show its dominance or back down. Both roosters and hens will also show their dominance by flaring their hackle feathers, which are located on their necks. Sometimes a bird will drop a wing and dance around in a circle to show the others who's dominant. This can all look funny to watch and a little violent, but humans should not interfere unless a bird is hurt. Usually this process looks worse than it actually is, and none of your birds will be injured.

The pecking order is ever changing, with lower-ranked birds challenging higher-ranked birds for a chance to move up. Within the order, it's not unusual to see friendships form. You'll often see hens broken off into friend groups that hang out together throughout the day. If a friend is lost or gets hurt and has to be removed from the flock to heal, her other friends can sometimes be seen standing in the spot they last saw her and looking for her. Once the bird returns, the friendship resumes.

Preening, Dust Bathing, and Sun Bathing

Preening is something that chickens do at least twice a day, and this behavior is easily spotted. You'll see a chicken standing still and rubbing her head along her tail and then along her feathers. What she's doing is gathering some preening oil from her uropygial gland, which is often called the preen gland, and distributing that oil through her feathers. In ducks, this oil keeps them waterproof as they swim. In chickens, this oil makes the feathers more water resistant and keeps them healthy, which means they last longer and are less likely to break.

Dust bathing is an essential chicken behavior and an opportune time to observe a flock's pecking order in action. If you've never seen a dust bath before, it can initially look like your chickens are dying. They are usually laying spread out in the dirt at ominous angles and sometimes they even look unconscious. On further inspection,

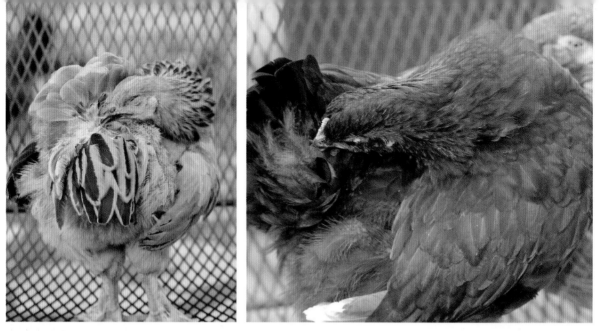

When a chicken preens you'll see it twist its head and neck to reach the preen gland. It will take oil from the gland and spread it through its feathers to clean them and help with waterproofing.

Dust bathing keeps a chicken clean and is a great time to observe social order in the flock. This rooster dust bathes after his hens have already had a turn. Quite the gentleman!

Sunbathing, even on the hottest days, can be relaxing—but it also serves a larger purpose. The heat from the sun encourages parasites to move to spots where they are more easily reached by the chicken.

the birds aren't dead or in the throes of dying. They are just so deep in enjoyment and relaxation that they are hard-pressed to respond.

Note

It's important to understand exactly why chickens dust bathe. By bathing, chickens are able to remove mites and other parasites as well as old skin and excess oil. This keeps them clean and healthy.

No matter where your dust bath is located, you'll notice that the dominant birds will bathe first. They will locate the best spot for the bath. Then they'll start to dig and move their bodies to clear out an impression big enough for them to fit. The impression will get bigger over time

as the birds work further into the hole, scraping and throwing dirt over their bodies and working it into their feathers. If others in the flock notice the dust bath, you can frequently see them marching around and around trying to join the bathers. If the bathers are the dominant birds, you'll find they won't leave the bath until they are good and ready. Conversely, if the bathers are less dominant birds, they'll have to leave since others outrank them. Sometimes less dominant birds will try to fit themselves into the hole. This gets easier as the hole gets bigger and there's more room. But until that point, the dominant birds will not move over and make room.

Sun bathing sometimes takes place while chickens are dust bathing, but other times you can walk into a chicken yard and see your chickens seemingly unconscious with their wings spread out as they lie in the hot sun. It can be an enormously hot day and when you see this behavior you start to wonder why? Why would chickens knowingly expose themselves to such hot temperatures?

Sun-bathing chickens are actually purposefully exposing certain parts of their skin to sunlight. Their sun-bathing postures give away what parts they are trying to expose.

- They may stand with their back to the sun and puff their head and back feathers.

- They may lie down and spread their wing feathers and tail feathers.

- They may lie down and turn over to expose areas underneath their wings and their breast.

Chickens are not the only birds that exhibit this behavior. You can see wild birds sun bathing too. In cold weather, sun bathing makes total sense. The birds are warming themselves without using a lot of valuable energy. If birds are wet, sun bathing is the perfect way to dry off. Sun bathing is also a valuable health tool. External parasites such as lice and mites can wreak havoc on a chicken's health, and these pests are not fond of overly hot and exposed conditions. So by purposefully sun bathing, chickens can encourage parasites to move to cooler locations that are easier for the chickens to reach and then pick off the parasites. Sun bathing also helps to warm a chicken's preening oil, making it easier to spread and distribute evenly through their feathers. This is why you often see chickens preening immediately following sun bathing. And the ultraviolet rays from the sun convert the chemical compounds in their preening oil into vitamin D which helps to maintain a chicken's good health.

Vocalizations

People that don't have chickens tend to think the only sound they make is that of a rooster crowing. Nothing could be further from the truth! Chickens have a language all their own, which they use quite often. What's more, their vocalizations start early.

While still in the egg, a mother hen will talk with her chicks through clucking sounds. She will offer comfort and encouragement. Once they're outside the egg, the chicks can recognize their mother hen and each other and start building their relationships.

Chicks will communicate among themselves and with their moms or their human caretakers. A brooder full of chicks is not a quiet place. If the chicks are content, they will happily scratch and peck and chirp to each other as they go about their business. If they get lost from their group, they will chirp loudly and with obvious distress. If they get cold, that chirping is just as loud and just as upset.

Mother hens talk with their babies quite a bit. When the chicks are in their eggs, mother hens will purr to their chicks. This helps the chicks recognize her when they hatch and tells them what's happening. A mother hen will cluck to her chicks when she's pointing out something good to eat. If there's something that's not good to eat, she'll vocally point that out too. Broody hens and mother hens also growl when their nests or chicks are disturbed.

Chickens have a language of their own. They are known to have at least 24 different calls.

FLOCK BEHAVIOR

Flock members will do the same thing as chicks: They will chirp back and forth to each other as they're grazing and going about their days. A hen will also sing an egg song after she finishes laying an egg. The level of singing can vary from breed to breed, but it will often provoke others to sing too. Some days an egg song can turn into an egg cacophony! No one is sure why hens sing an egg song. Some speculate they're proud of their laying accomplishment, others say they want the rest of the flock to know where they are, and some say it's a way of distracting predators from the nest as the hen is moving away from an egg she just laid.

If a rooster is in a flock, he will sound different alarm calls for different types of danger. The same is true of the lead hen of a flock with no rooster. Often the alarm call for an aerial predator is much more high and shrill than the alarm call for a predator on the ground.

Chickens also make growling types of sounds when they're frustrated, such as if they need more food or they just can't wait for a hen to leave the nest box they want. They will also make high-pitched sounds of encouragement, like when you bring treats to them and don't give them out soon enough.

Listening to your flock's different vocalizations is fun, and it's a great way to get to know them and bond with them. Soon you'll understand some of their language and be able to "talk" right along with them.

Roosting

Roosting is another chicken activity that shows the pecking order in action. Roosting is an

Roosting is a natural instinct to keep safe at night. It's also a clear indicator of the social structure in a flock. The higher order birds get their choice of spots.

essential survival tool for birds that are ultimately prey animals. After all, the higher you can get and the more you're buffered on both sides, the better your chances for surviving the night.

Many chicken coops have roosting bars at multiple heights while others have just one long bar that runs the length of the coop. Either way, there are preferred spots, and friends like to be next to each other. Usually around 30 minutes or so before the sun sets you'll see your chickens filing into the coop for the night. This is when the action takes place. The dominant bird takes the best place on the roosting bar, and she will defend her spot. As others file in, she will peck and squawk to move birds around to her satisfaction. If by chance a lesser bird has gone to the coop a little earlier to land the most coveted spot, that bird will be unceremoniously moved.

Troubleshooting Relationships

Sometimes in life people don't get along, and sometimes birds don't get along either. Just like humans have to cope with this, chickens do too. The difference is, some of your flock may need a little help from you in managing their relationships.

"Not getting along" can take different forms. One bird may be a little more aggressive than the others. Usually that bird is at the top of the pecking order and trouble happens when she's trying to get her share of the food first or jockeying for the best roosting bar position. She will try to keep others away from what she wants. She might squawk, lunge at, or chase the others. She means business when her lower-ranked companions aren't behaving the way she wants them to behave.

By now, you know that this type of behavior does happen naturally as part of establishing and keeping the pecking order. But what's key is to closely monitor the health of your birds. Sometimes this behavior leads to bullying, which can lead to some birds in your flock being deprived of proper food and water on a daily basis. In this case, you'll want to set up more food and water stations so that everyone has an equal chance at good nutrition. After all, it's hard for a bully to patrol more than one food and water station at a time.

Feather loss is one good sign that you have a bully. True, feather loss in itself isn't a problem since feathers do grow back. But that bare spot can become a problem when it's pecked. Chickens are curious and they like to explore things that look different. That exposed skin looks different, and it can easily be damaged and bleed. When that happens, chickens can be relentless. They will keep pecking and pecking, causing severe injury and sometimes death. An injured bird should be separated immediately from your flock and should only be returned to the flock when the injury is treated and covered or has healed. There is also a type of pecking called aggressive pecking, which is meant to cause injury. This type of pecking occurs exclusively at the top of the head or comb.

COMMON CAUSES OF FEATHER LOSS

- Feather picking associated with bullying.

- Frequent rubbing can wear feathers in certain places as birds try to fit in small areas. For example, birds can rub off feathers as they try to squeeze in and out of a fence. Hens can have feather loss on their necks and tails as they duck in and out of nest boxes.

- Treading during mating (see page 61) causes feather loss on a hen's back and sides. There can also be feather loss on the head as the rooster gains balance and hold.

- Broody hens will remove feathers from their breast to line their nest.

- Feather loss from molting occurs as baby chicks mature and yearly in mature chickens. (See page 155.)

Preventing & Stopping Bullying

Chickens are busy creatures that ideally spend the majority of their day foraging. When they don't have access to outdoors and free range, they get bored and start looking for things to do. Boredom leads to bullying. But there are some easy boredom busters that will keep your birds too busy to bully.

BOREDOM BUSTERS

- Hang a whole cabbage from the ceiling just high enough for the chickens to reach.

- Cut a watermelon or squash in half and place it in the coop.

- Clean the coop and leave the bale of straw or bag of chips intact but without strings or wrapping attached. Let the chickens spread it.

- Give the chickens a flock block to eat. Set the flock block in a shallow dish planter to keep crumbs contained.

- Add lots of branches at varying heights to the run.

- Add a chicken swing.

STOPPING A BULLY IN HER TRACKS

Bullies that become a real problem should be completely removed from the flock for a week. During that time the flock should not have access to the bully. You can keep your bully in the garage or a safe barn inside a large dog crate with lots of bedding and food and water. This will give your flock time to establish a new pecking order without the bully. Once the bully returns, he or she will have to re-establish a place in the order; hopefully that place will be a little further down in ranking.

Flock blocks are a nutritious and fun treat that also help fight boredom and keep chickens occupied.

If a simple removal and return of the bully doesn't work, then you can try something that I call the home aquarium method. If you've ever had aquarium fish, you know that when you introduce new fish, it's best to rearrange all the hardscape in the aquarium so all the fish have to establish new territories, not just the new fish. This same method works for chickens, too. First remove the bully bird. Then rearrange the coop as much as you can. Move the nest boxes and roosting bars to a different spot. If you keep food and water in the coop, move that, too. Then let the flock, minus the bully, settle in for a few days. While the bully is gone, the birds will have to rearrange their pecking order. Once the bully returns, he or she can't just go back to the same old routine: the bully's favorite nest boxes and roosting spots are all gone. Looks like the bully will have to find a new spot in the pecking order! This method can be used when you're introducing birds to the flock, too. You'll want to completely integrate the birds at night. Let them out of the coop for the day, and, while they're gone, rearrange everything. That throws everyone into a new pecking order and gives the new chickens an equal chance to integrate.

Tip

When you're rearranging the coop, you'll have to be careful to keep nest boxes open for birds that need to lay eggs throughout the day. Also, you may notice a drop in egg production because of the stress of rearranging the coop and pecking order.

Piling

Piling is a self-descriptive problem; the birds pile on top of each other and suffocate the birds that are underneath the pile. This is an especially frustrating problem because in some cases you are powerless to stop it. (Luckily, in all my years of chicken keeping there was only one flock where I had a problem with this. Hopefully you will also be fortunate and experience minimal piling!)

Piling can occur with baby chicks in a brooder. If the heat is too low, they'll pile to keep warm. If the heat is too high, they'll pile to get away from it. Monitoring the heat is essential and keeping a brooder with rounded sides can help.

Piling also occurs when birds are frightened. For me, this happened when my chickens first moved to an outdoor coop. During the night, we had a terrible thunderstorm. In the morning I found my birds, flattened and dead on the floor. Other frightening events that can lead to piling include predators trying to access the coop and sudden power loss in a coop that's usually well-lit.

Piling behavior happens most with younger chickens at night and usually before the birds are all settled. So make sure your birds are roosted before the coop is shut; separating them if you have to and making sure they get on the roost bars. Try to keep predators at bay, of course. And, if you experience a loud event like fireworks or a thunderstorm, check your birds as soon as you can safely do so. Make sure they stay on or are put back on the roost bars and that they aren't milling about the floor of the coop. Once the birds are older, this behavior is not as common.

Q&A: Chicken Behavior

Can you introduce younger chicks to older chicks?

This question gets back to one that scientists have been studying. When does the pecking order form in groups of baby chicks? Since the pecking order is normally established around 8–10 weeks of age, you've got a window where you can introduce chicks to chicks without any special attempts at getting them used to each other. I have done this before with the chicks being about a week apart in age, and it worked perfectly. A good rule of thumb is under five weeks old you can go ahead and add them. After that, you'll need to set them up in adjoining areas so they can get used to each other just like the process for introducing adult chickens.

I have breeds of chickens that are supposed to be friendly and they're not. Then I have breeds of chickens that aren't supposed to be friendly and they are. Why is this?

As with people, each chicken has a distinct personality no matter its breed. This can be true even if two chickens are raised together and in the same way. For instance, I have two Buff Orpington hens. They are supposed to be a friendly breed. One is just that, she's very friendly and loves to be petted. The other is awful. She pecks at people and never gets petted. On the other hand, I've had White Leghorns, and they're known to be flighty and not friendly. But I find their personality to be wonderful and friendly. I've even had one that curls up in my lap and falls asleep.

Can chickens understand what we say?

I think chickens get used to us and whether we know it or not, we are repetitive. We say things in the same tone and cadence over and over again. They start to associate our sounds with actions. Probably the better question is can we understand our chickens? I think we can if we pay enough attention. I've noticed that when I hold a chicken and talk directly to it, the chicken will look at me and make a series of sounds. I can imitate those sounds back to the

Working with Your Flock

It's important that you have a working relationship with your flock. Some people call this training, though for me that brings up images of chickens being trained to run obstacle courses for the county fair! The truth, however, is that training for a backyard flock is actually utilitarian. The goal is to master simple techniques to get your chickens where they need to be, when they need to be there. In an emergency situation, it can literally be a lifesaver.

If you're wondering how you can train a chicken, it's a matter of understanding how

These teenage chicks are checking out the adult coop where they'll soon by living.

chicken and it will make the sounds again. This has happened with different chickens through the years, so it's not a coincidence. I can't be sure of what my chickens are saying to me, but I know it's not bad since they'll sit and "talk" with me. I wonder, is it my name? Or are they telling me about their day? Their tone and cadence is always the same so it has some meaning. I just have to figure it out!

chickens communicate and how we communicate with them. Always keep in mind that chickens are visual and they're verbal, plus they like food. Drawing on a chicken's natural behavior and instincts, you'd be surprised how well you can interact with your flock. We know chickens are flock animals. They interact together all day and stay together as a group to keep safe from predators. So you need to be seen as a member of their flock—and hopefully one that's high in the pecking order. When working with your chickens, it's also important to find a routine and use it daily. The repetition will reinforce the behavior. Voice training and stick training are two techniques that are simple and effective to use every day.

Voice Training

I have kept the same routine with all my flocks. This starts when my chickens are baby chicks in the brooder. I give them the same greeting each time I visit them. I say, "Hey pretty ladies," and then I talk to them during our time together. I also like to put some food in my hand and let them eat out of it. They get to know me and I get to know them.

As the chicks grow and move to the backyard, I keep up the same routine. I greet them the same way each day. (It's the same greeting I used with them as baby chicks.) When I give them treats, I use the same wording and cadence to call them.

Even if they've seen me and are already heading toward me, I still call out to them. I always say, "Here chickens, here chickens. I've got a treat for you!" I also use the same bucket each time I carry out treats. There may be other containers that actually carry the treat, but I always bring the bucket too. This adds another level of familiarity. The chickens associate the bucket with a treat.

This is similar to the way chickens communicate with each other. Think about a rooster. When he finds a great treat to share with his hens, he vocalizes so the hens hear him and know to join him. He uses the same vocalization every time. Chickens are smart! You'll find they understand

You'll know your voice training is successful when your chickens come running to you even if you don't have their treat bucket.

Stick training is an easy way to move and guide your flock without a lot of drama.

our repetitive behavior in the same way if it affects them. Repetition reinforces their learning.

When adopting older chickens, this technique still works. If you already have a flock in place and you're adding to it, the adopted chickens will learn quickly by observing how the existing flock interacts with you. They'll simply join the flock's routine. If the adopted chickens are your only flock, then just start with this type of routine from day one. The chickens will soon see you as a trusted member of the flock.

If you want to train your chickens for obstacle courses and other fun tricks, remember that it's not as much about the food treat; it's about consistency when communicating. You can use verbal, visual, and food affirmations to help your chickens respond in the right way.

So, why is training a chicken to come to you important? There are times you'll need to gather your chickens. You may be going out for the afternoon, but your chickens have been free ranging and you want to put them away. Or maybe your chickens have gotten out of their safe area and need to be called back. Also, what if you've had an emergency? If a predator has attacked and your

hens have scattered, you'll be thankful you can call out to them and get a response.

Stick Training

After you've used the voice training technique to call your flock to you, you may need to move your chickens around, say to put them safely away in the coop.

Try using a stick to move your flock. It doesn't have to be a fancy stick, just something nice-sized that you can easily hold horizontally from your body. I've used fallen branches from trees and even an old metal broom handle. The metal handle may not look pretty, but I've found it holds up well.

If you've been to a fair livestock showing, you may recognize this technique. To start the training, hold your arms horizontally out to the sides with the stick in one of your hands. This makes you seem very big. Then move slowly toward your flock to guide them. They will start to move in the direction you guide them. If they need a little encouragement, you can rap the stick on the ground. At no time do you hit your chickens! The stick is just a guide post.

The first time you try this, your chickens may scatter in every direction. But be patient and keep using this technique. Your birds will get used to it and you'll soon be easily guiding your flock with your stick. In fact, sometimes you may find you don't need the stick. Your arms held out to the sides may suffice.

How to Safely Integrate Old and New

Once you've decided to expand your flock, integration looms large. You've got an existing flock and a beautiful new set of birds. Will they ever get along? Yes, they can get along, but integrating flocks can be a stressful time for both you and your birds. Think of it this way: Your flock has been living in harmony with an established pecking order and routine. Now a bunch of newcomers are introduced and the existing birds don't know whether they are friend or foe. They've got to make acquaintances and everyone has to re-establish their place in the pecking order. Don't worry; there are ways to make this easier on everyone!

First, make sure the coop where everyone will ultimately end up is big enough to handle the increased flock size, with enough floor space, nest boxes, and roosting space. Make sure you have some extra food and water stations on hand so there are plenty of places to eat and drink. This can all be done while your new birds are growing or being quarantined.

If you're integrating new chicks that you've hand raised, introductions can start early as you take them outside, weather permitting, for some time in the fresh air and sun. They should be separated from the existing flock, but they should still be able to see them. You'll notice your older chickens are interested in their new flock mates. They'll come over and check things out. The younger chicks will do the same. And since they can only see but not touch, these are safe interactions at an important time. Think of them as non-verbal conversations where the chickens start to work out the pecking order.

Separate outdoor time can go on for a while, and it's a good rule of thumb to let your young chickens get as big as possible before completely integrating them. Complete integration is when everyone is living together all day every day, making one whole flock. It's important that the young chickens are around the same size as the mature birds when this time comes, as the new birds need to be able to hold their own as they work to establish their place in the flock. Birds that are too small run the chance of being bullied or hurt. Generally, complete integration should be no sooner than when your new chickens are six weeks old, but I recommend waiting longer. I've found waiting until the new chickens are 12 or more weeks old increases your chances for a stress-free integration. The exception to this rule is if a broody hen is making the introductions. She knows when and how to make this successful and will be there to support her brood as they integrate into the flock.

In the weeks leading up to the integration, you'll need completely separate housing so your two flocks each have a safe place to eat, drink, and sleep. I usually set up a temporary night coop

This adult Buff Orpington (left) is sharing a watermelon with Speckled Sussex and Buff Brahma youngsters. Sharing food is a way for adults and young chickens to get to know each other and establish a rapport.

in my garage and have a temporary run outside during the day. My young birds use the temporary housing while my older birds get to stay in their normal space with their normal routine. As the birds get older, you can use your judgment and even let them roam freely together outside during the day, separating the flocks only at night. This makes sure everyone has a chance to have a safe place to rest, but at the same time lets integration begin. Moving gradually is the key here.

Once the new chickens are big enough and confident enough for complete integration, then it's common practice to integrate the flocks at night. To do this, first let the existing group get settled in their coop and go to sleep. Under the darkness of night, with minimal help from a flashlight, take the new girls from their temporary home and gently place them on the roosting bars in the main coop. Enlist the help

of a family member if you can so that you can do this as quickly as possible. If you create very little commotion, everyone usually settles in and goes back to sleep for the night. Of course, it's a good idea to wait in the dark just outside the coop until all is quiet and everyone's settled. Early in the morning, open the coop and let everyone out.

Free-range birds often have an easier time with integration because they are more focused on getting outdoors and finding grass and goodies. If your birds don't free range, it's important to give everyone lots of hiding places so they can get out of the fray. A good idea with cooped chickens is to keep them occupied and get their minds off the newcomers. Use your boredom busters (page 44)! Anything to redirect their energies works well. Soon you'll find that your old flock and new flock are working together with a peaceful new pecking order.

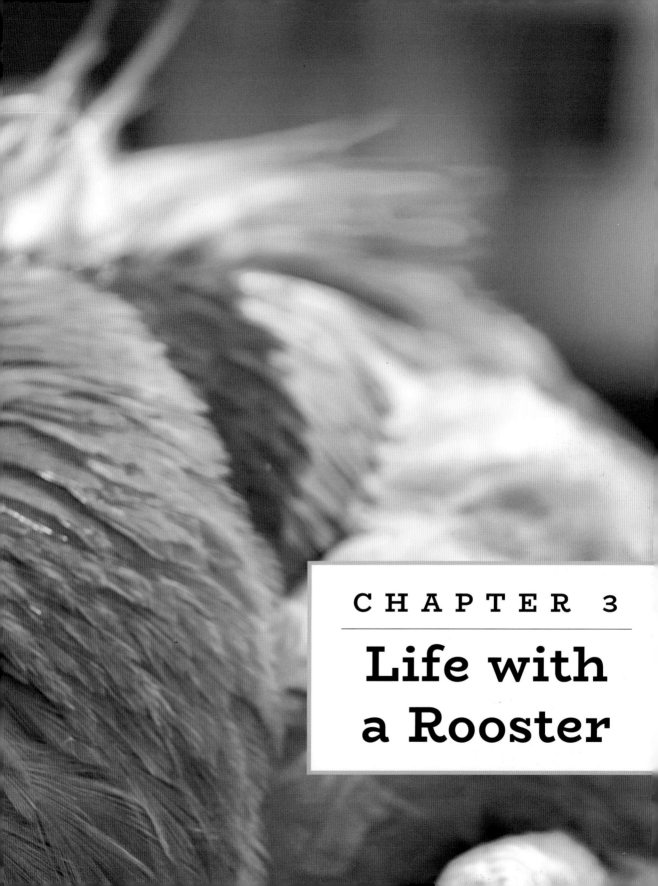

CHAPTER 3
Life with a Rooster

54

One year when we were buying new chicks to expand our flock, I was at the feed store picking out chicks. I had asked to pick out my own chicks and was picking from the Buff Orpington brooder marked as a pullet bin. As I picked up the last chick and gave it an inspection, I noticed it had droppings on its wings. This didn't bother me much, but the feed store person suggested I pick another chick. So I sat the chick down and looked for the biggest, healthiest looking chick in the brooder. One stood out above the rest, so into the box it went. We arrived home and the family named the new chick Kate—my kids fell in love with her.

Over the next few weeks I noticed Kate was growing faster than everyone else. Also, her comb and wattle were a deeper red and they were bigger than the rest of the birds. I hoped against hope that I was mistaken. Yet when Kate and the other chicks were in the temporary garage pen, my husband and I received the confirmation loud and clear. As we worked filling feeders and waterers, the sound rang out that fateful day: a choked and raspy sounding cock-a-doodle-do.

We both stopped what we were doing and looked at each other. We didn't say a word. We were both silently hoping we wouldn't hear it again. Oh, but we did. Soon after the first, there was a second pathetic attempt at a cock-a-doodle-do. The crowing soon became a daily occurrence and got louder and louder, which is normal as roosters mature. We couldn't deny it anymore; Kate wasn't a Kate. We transferred that name to our other Buff Orpington and renamed our beautiful Buff Orpington rooster Roopert.

I was initially upset by this turn of events. I had no idea what to expect out of a rooster. However, while the dynamics in our flock did change, I soon found it wasn't the end of the world. I found that watching a rooster mature was also a magnificent thing to behold. They are so much bigger birds than the hens and their coloring is so bright and vivid. The extra layer of protection a rooster offers also lends comfort as the flock free ranges. And of course, rooster antics up the backyard entertainment value. All in all, having a rooster in our flock has been a positive experience. I hope it will be the same for you!

Identifying Hens and Roosters

Statistics show that 50 percent or more of a clutch will be roosters. If you're hatching eggs through a broody or an incubator, you will definitely get a few roosters. If you're buying chicks, on the other hand, you will have a much smaller chance of getting a rooster. Hatcheries are good at sexing chicks as most spend lots of money on a good chicken sexing expert—many claim to be 90 percent accurate. (Sexing experts actually make a nice salary after they've gone through a six-month trade program and apprenticed for two years!) Even with all this, the reality is that no matter which way you expand your flock, unless you're extremely lucky, you'll end up with a rooster at some point.

For some chicken keepers, this is not a problem. They go ahead and raise the chicks to maturity and then eat the roosters. Others raise

This young rooster's large and deep red comb and wattle are the first indications of his sex.

only rare breeds so their roosters are highly desired—making for an easy sale. However, if you live in an area that restricts roosters and you're squeamish about eating your birds, then you need to have plans made just in case your luck runs out.

If getting a rooster is a big problem for you, a good option for avoiding them is to buy day-old sex-linked chicks. Sex-linked chickens are cross-bred, so at hatch you can tell their sex by their color. They are known as good egg producers with good personalities. They go by different names at different hatcheries. Most commonly, red sex-links are a cross between a Rhode Island Red rooster or a New Hampshire rooster with a White Plymouth Rock, Rhode Island White, Silver Laced Wyandotte, or Delaware hen. They are also known as Cherry Eggers, Cinnamon Queens, Golden Buff, Golden Comets, and Red Stars. Black sex-links, also known as Black Stars,

are most often a cross between a Rhode Island Red Rooster or a New Hampshire Rooster with a Barred Plymouth Rock hen. ISA Browns are a popular sex-link created by crossing a Rhode Island Red rooster with a White Leghorn hen. There is also a white-colored sex-link chicken known as the California White.

> ## Fun Fact
>
> Sex-linked chickens are not an actual chicken breed; they are a crossbreed also known as a hybrid. A breed is defined as a group of animals with the same characteristic that predictably reproduce that characteristic when they are bred together. Sex-linked chickens do not breed true. They can breed and reproduce but their characteristics are not guaranteed to show up or remain consistent.

Sexing

If your chickens are not sex-linked, then you're left wondering exactly what you've got. I hate to say it, but many times there is no definitive way to tell until your chickens actually lay eggs. First let's run through some fun folk tales that people like to use.

• Lowering the incubator temperature by half a degree results in hens and raising it by half a degree results in roosters.

• Tie a heavy object on a string or thread a sewing needle and hold it above a chick. On its own, it will circle if you have a hen and it will move back and forth in a straight line if you have a rooster.

Feather development can be a helpful indicator of the sex of your chickens. Roosters have longer pointed hackle feathers around their neck. Hens have shorter, more rounded hackle feathers.

- Rounded hatching eggs are hens and pointy-shaped eggs are roosters.

- If you wave a hat above your chicks, then drop it, the hens will squat or run but the roosters will stand at alert.

- Rub a penny along the back of a chick; then toss it. If it comes up heads you have a rooster. Tails equals a hen.

- Two days after hatch rub your finger on the chick's belly near the back vent. If there's a hard lump it's a rooster. No lump equals a hen.

All these tales are fun and a great way to involve your kids in chicken keeping, but they are not accurate. Out of all of them, the hard lump on the belly of a rooster comes the closest to real life accuracy. Professional vent sexers will squeeze the feces out of a chick and open the cloaca to see if it has the telltale small bump of a rooster. However, it's important not to try the professional method on your own chicks because you can easily injure them!

Fun Fact

Chickens start to develop their sex organs around days five to seven of incubation in the egg. Sex is determined by chromosomes coming from the hen. This is the opposite of humans where sex is determined by chromosomes coming from the male.

Wing sexing is another technique that many people use. It's done only within the first 48 hours of hatch. A hen will have two different feather lengths on its wing while a rooster's wing feathers will be the same length. In some strains within some breeds this can be very accurate but the reality is most of us don't have those particular breeds and strains in our backyards.

It's best to watch for general behavioral and physical differences as your chicks grow to see if you have a rooster or a hen. Here are some things to observe:

• Roosters will be larger than hens.

• The combs and wattles on a rooster will be darker pink. They will grow faster and larger than a hen's.

• Roosters will be more bold and friendly early on while hens are less friendly. (This changes as the birds grow.)

• If you have more than one rooster, they may chest bump and challenge each other.

• Hackle feathers are located on the neck. In roosters those feathers are longer, more pointed and narrower. Hackle feathers in hens are more rounded and oval shaped.

• Roosters have long, skinny saddle feathers located where the back meets the tail. They start to develop around 12 weeks of age. Hens have rounded feathers and don't have the long, skinny saddle feathers.

• Roosters will switch from chirping to crowing around four months of age.

• Roosters will have thicker legs and may develop spurs early on. They are larger than a hen's.

While these are good general guidelines, they are not 100 percent accurate. Sometimes hens will have very large combs and wattles. Sometimes they will crow. Hens can be extremely friendly and they will chest bump each other to establish the pecking order. Feather development is one of the best indicators of sex and it's the quickest since feathers develop around 3 months of age; however it does not work in Silkies or Sebrights. The ultimate test is whether or not your bird lays an egg.

Flock Dynamics with a Rooster

If you've never had a rooster in your flock, you'll find the dynamics change when a rooster is added. Hens that looked up to you and squatted as you

A good rooster is always watching over his flock, protecting them from predators, finding food, and helping to mediate squabbles.

walked in the yard will no longer pay as much attention to you. They now have a rooster instead of a human leader. But it's not all bad.

People make roosters out to be horrible, yet there are actually pros and cons. Let's take a look at both.

PROS

Protection: When your flock is without a rooster, usually there's a lead hen that will take over guard duty. However, there's nothing like a good rooster. You'll find that a rooster who takes protection seriously will always have an eye or ear on alert for predators. If something is near that a rooster doesn't like, he will gather his hens in a safe, protected spot. If some of his hens have wandered and he's not close to them, he will sound an alarm by calling loudly. Once you've heard this call, you'll recognize it every time. If all else fails, he will fight even to the death to protect his hens.

> ## Tip
> If you have more than one rooster, make sure the steadfast protector is allowed to lead the flock. Sometimes this means separating roosters, but in the end it's better for your flock to have good protection. Just don't always assume the biggest, prettiest rooster is going to be your best protector. That's not always the case.

Mediation: We talked about chickens not getting along in the last chapter. Well, roosters are great to have around when hens squabble! They don't like discord within the ranks, so they will mediate disagreements between the flock members and keep the peace in general.

Reproduction: If you don't want to rely on hatcheries, you've got to have fertilized eggs to expand your flock. That means you've got to have a rooster if growing your flock size is a goal. Also, in flocks that have roosters, the pullets tend to mature and lay eggs faster than in flocks that don't have roosters.

Foraging: Roosters are all about their genetic destiny and the only way to ensure that legacy is by having a group of hens healthy enough to lay eggs and raise the young. This means roosters will spend large amounts of time ferreting out the best treats in your coop or yard. They will then signal the hens that they've found something and let the hens eat first.

CONS

Noise: Roosters are loud and they don't just crow in the morning. They will crow to each other if you have more than one. They will crow if they feel threatened or hear other noises. They will crow all day! Noise is definitely a consideration.

Neighbors and Legal Issues: Many neighborhoods don't allow roosters. Even if they do allow them, it's always a good idea to consider your neighbors. While you may not find a rooster crowing offensive or disruptive, others might. So make sure everyone within earshot is on board before adding a rooster.

Fertilized Eggs: A fertilized egg has the ability to become a chicken, but a chicken is not formed until the egg is incubated. I can tell you fertilized eggs don't taste different and they don't have a chicken inside. Still, some people will not eat fertilized eggs.

Aggressive Behavior: Some roosters can become highly aggressive and can cause injury to pets and people. This behavior doesn't normally present itself until roosters mature and their hormones are raging. You may be able to tame this aggressive behavior by showing your rooster who is boss. But if not, it's good to have plans for your mean rooster. Many people will choose to re-home their roosters or eat them.

Mating

Before we talk about the actual mating process, it's important to understand a rooster's reproductive system. Roosters do not have a penis or external reproductive organs. Their sperm are viable at body temperature, unlike humans, so everything is housed in their body. Roosters have two bean shaped testes, which are located in front of their kidneys. The testes produce testosterone and sperm on a regular basis. Since a rooster's reproductive system is influenced by lighting just like a hen's system, the testes shrink and grow seasonally. Deferent ducts carry sperm on a trip that takes one to four days, to a storage area in the cloaca where there is a small papilla, or bump, that is the mating organ.

When chickens mate it's quite a sight, both comical as the rooster dances and violent looking

A large red comb with tall points is attractive to hens. Since combs can be an indicator of good health, this makes sense.

during the actual mating. Roosters spend their days trying to impress their hens; they'll tidbit to show hens the best treats in the yard and find the ideal sun bathing spots for the hens. If a rooster wants to mate, he will woo his hen of choice by lowering his head and puffing all his beautiful decorative feathers. (This is the reason for those hackle and saddle feathers.) Then he drops a wing and dances with little hops back and forth.

Hens are very specific when they're looking for a mate. Here are the qualities they find most attractive. They look for large, bright red combs with tall points, evenly formed wattles and big spurs. This makes sense because these are all indications of a healthy bird that can carry on a strong genetic line. Interestingly, feather color matters little.

If a hen is receptive she will crouch or squat to let the rooster know. The rooster grabs her head and neck with his beak and then stands on her back. He uses a walking or shuffling motion, called treading, on her back to maintain balance. The rooster sweeps his tail to move her tail feathers to the side. Since roosters don't have external sex organs, the act of mating is just the touching of cloaca, sometimes referred to as a cloacal kiss. At this time, the hen everts her cloaca so the rooster's sperm has a better chance of making its way up her oviduct.

Fun Fact

Even in a flock without a rooster, hens will crouch or squat to indicate they are ready to mate and are submissive. This often happens when you walk up to a hen in the chicken yard. And with young hens, it's a good indicator they are mature enough to lay eggs.

Mating itself doesn't hurt a hen, but hens can suffer injuries from the process. Some roosters are a little too heavy for the hen and this can cause leg injuries. A rooster's spurs can cause cuts on a hen's back and sides as well. A common injury is the loss of back and wing feathers from treading.

To make sure your hens don't get injured from too much of a rooster's attention, start by maintaining a safe ratio. Ten to twelve hens to one rooster is a good ratio to maintain, on average. The goal here is to have enough hens so a rooster's attention is distributed evenly and no one hen is hurt being overly mated. Yet even with enough hens, I find that roosters have their favorites. You'll soon find certain chickens with their backs or wing areas becoming bare. You can fit those hens with a saddle to prevent further damage and let feathers grow back unmolested.

Tip

Hens may not like their saddle at first. Some may not tolerate a saddle at all. However, I find most get used to it after a day or so and then resume normal activities.

All About Spurs

Spurs are associated with roosters and they're used for protection and fighting. They can be intimidating and roosters that aren't well behaved use them when they chase humans and even other pets away from the chicken coop. Spurs can cause real damage, so they should be respected.

Spurs are an extension of the leg bone that's covered with a hard layer made of keratin, the same substance that makes up our fingernails and hair. Both hens and roosters start out life with a spur bud that has the potential for growing spurs. In some breeds it is expected that both sexes will grow spurs; Polish hens are known for growing spurs as well as Mediterranean breeds such as the Leghorn, Minorca, Sicilian Buttercup, and

Spurs help a rooster defend himself from predators and from other roosters
that pick a fight as they look to climb the social ladder.

Ancona. But the reality is that hens of any breed can grow spurs. It can happen as early as three months old, but usually doesn't happen until the hens are older.

You'll want to keep an eye on your rooster's spurs as they can grow too long. They may become a hindrance when the bird walks and they can cause deep cuts when they mate with hens. Spurs can also curl as they grow and reach the leg, cutting it and causing lameness or infection. This is not as likely for hens since their spurs are generally shorter. However, they should be watched as well.

If none of these scenarios take place, then no maintenance is required.

Spur Maintenance

If you do need to perform spur maintenance, there are several methods available—clipping and filing are my preferred methods. All these methods are easiest done with two people, one to hold the rooster and one to perform the task. The temperament of your rooster determines the resistance you'll meet. (My rooster actually has no problem with maintenance and will fall asleep while he's being held.)

Clipping: Rooster spurs can be clipped just as you would trim nails for a dog or cat. You need to hold the rooster tightly though so he doesn't move. You don't want to make a hasty cut as the rooster is flailing. You also need to be absolutely sure of where the soft bone is located. If you hit the soft bone, it can bleed and be very painful. Make sure you have cornstarch or styptic powder

on hand and some wound spray just in case. It's good to use a sharp-bladed clipper so the cut is quick and painless. If you don't have a sharp blade or a big enough clipper, the spur can split. This can cause damage and be painful. Out of an abundance of caution, I like to clip small bits at a time.

Filing: This is a little safer than clipping since you're removing small bits of spur as you file. You can use a hand-held file or a rotary tool for this. You'll still need to secure your rooster. You'll also still want to have cornstarch or styptic powder and an antiseptic spray on hand. File slowly and watch out for the soft bone. Since you're not removing large amounts, you may want to do this more often. I usually clip and then switch to the file as I get closer to the soft bone. The file is also great for smoothing any jagged edges.

Uncapping: This method removes the entire outer hard shell of the spur with a pair of pliers. Once it's done, the rooster will be left with a soft, smaller spur that will gradually harden. Some people gently twist the spur back and forth until it pops off. Others like to soften the spur and then twist it. To do this, heat a potato in the microwave until it's too hot to handle with your bare hands. The exact time will vary by potato size, but in general it takes about five to eight minutes. Carefully pick up the potato using something like an oven glove to protect your hand. Put the rooster's spur in the hot potato, making sure you leave room between the potato and the rooster's leg so that he doesn't get burned. Keep the potato in place and let it start to cool.

Q&A: Roosters

Do you need a rooster to have eggs?

Don't laugh if you know the answer! This is one of the most commonly asked questions in the chicken-keeping world and outside it. While this isn't a beginner book, the answer always bears repeating: no. You do not need a rooster for eggs. You only need hens. However, you do need a rooster for fertilized eggs.

What percentage of eggs will be fertilized if I have a rooster?

Roosters are pretty good at spreading their genetic destiny. You may not always see them mating, but they are busy. Chances are most of your eggs are fertilized if you have a rooster in your flock.

When do roosters mature enough to fertilize eggs?

Roosters are sexually mature when they are around five to six months old. Hens are sexually mature when they begin laying eggs.

Can I have more than one rooster in my flock?

Yes. It's easiest to have multiple roosters if they all grow up together or if you add them in a mixed sex group of chicks. Roosters will generally challenge each other and back down before injury. But this isn't always the case so you should watch your roosters closely and separate them if necessary.

Can roosters mate with their offspring?

Yes. A rooster can mate with any of his flock members. If you're eating the eggs and not using your rooster for flock expansion, then this isn't a problem. However, if you're raising breeding stock then you'll want to consider separation of related birds.

Remove the potato and use a pair of pliers to gently twist the softened spur to remove it. If it doesn't twist easily, you can reheat the potato and try again. The small spur bud that's left after uncapping will be soft and sensitive. You can separate your rooster from the flock until the spur starts to harden if you find he's having trouble with pain.

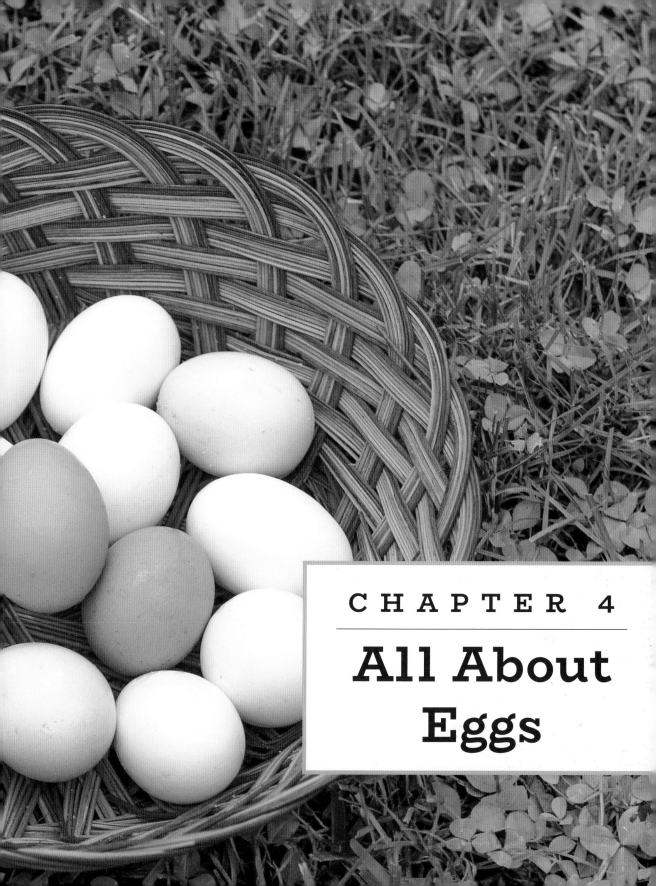

All About Eggs

CHAPTER 4

Once your chickens are finally old enough to lay eggs, it's exciting to head out to the coop and check the nest boxes to see if you've gotten any eggs that day. Your days are filled with anticipation!

When the first egg came from my first flock, I was so excited. I'd been waiting and checking the nest boxes every day. My hens had started squatting as I walked up to them, so I knew eggs were imminent. And then suddenly there it was! I remember grabbing that egg and running inside to show my family. We all stood around the egg and admired it.

We debated whether to keep that egg or cook it. In the end, we couldn't resist tasting it. So I took a picture for posterity's sake, then we cooked it and we all had one bite. It was delicious and we were forever hooked on the creamy flavor of the backyard egg.

Over the next few days, we received more and more eggs. I gathered up a dozen and decided the extras should go to the rest of our family. When my dad came to pick some up, I immediately opened the egg carton and preened over the beauty of my fresh eggs. My dad, who grew up in the city, just stood and looked at the eggs. I asked him what was wrong. He asked why they were smaller than store bought eggs. I explained that chickens lay small eggs when they first start laying but not to worry because they get bigger over time. He looked at me and said that was fascinating. He took the eggs home but looked befuddled as he was leaving.

I got a call from my mom that night. She asked what I had told my dad about those eggs. Apparently he had taken the eggs to my mom and told her not to worry that they were small, to just leave them in the refrigerator for a few days and they'd get bigger. He told her he never knew eggs grew!

It took me a second to realize what my mom was telling me. My dad thought that eggs themselves grew in size once they were laid until they reached the size you see in the stores. My mom was laughing so hard she almost cried. She explained to my dad that when an egg is laid, it doesn't grow in size. Hens that are just getting into the swing of laying, lay smaller eggs until their bodies fully adjust to the laying cycle. My dad seemed relieved by this information and has remained fascinated by my chickens. He loves the eggs they produce.

Beyond loving the eggs backyard chickens produce, this chapter takes fresh eggs to another level of understanding by exploring exactly how eggs are formed inside a hen, how eggs get their color, which chicken breeds lay colored eggs, the parts of an egg, and how to properly store eggs. This chapter will take the ordinary egg and make it extraordinary. You'll never look at an egg the same way again!

Egg Formation and Egg Color

Nowadays, as more people get into backyard chicken keeping, interest in the different egg colors and the breeds that lay them has increased. Many folks raise different breeds of chickens just for the color of eggs they lay! There's nothing wrong with this. In fact, it's downright fun to collect an egg basket full of colored eggs. There are so many egg colors to choose from: white, brown, olive green, dark chocolate brown, and blue.

Just how eggs get these unusual colors and whether or not they taste different are questions many chicken keepers have. We've all heard people say brown eggs taste better than white eggs. We've also seen people look at green and blue eggs and ask how they taste. I actually even hear it from chicken keepers themselves: they will swear that one color egg tastes different than another. They associate egg taste with shell color.

To understand how an egg is formed is to understand how egg color is applied—and why it

This Brown Leghorn's earlobes indicate the color eggs she lays.

doesn't affect taste. From this, you can see that the flavorful parts of an egg are intact before the eggshell is ever formed. Taste comes from what a hen eats and the freshness of the egg.

> **Tip**
> With some exceptions, you can generally tell what color egg a chicken will lay by looking at its earlobes. White lobes equal white eggs.

How Eggs are Formed

When a female chick hatches, she already has fully-formed ovaries, which contain all the reproductive cells (called ova) she will ever need to lay eggs. She has tens of thousands of ova contained in her ovaries. As she grows, her left ovary matures and becomes her only functional ovary. Each ovum has the potential to form an egg and is contained within a follicle, which is attached to the ovary. Once a hen is mature, she has the capability to begin laying eggs. At maturity, her body starts to add yolk to the ova. You can think of it like a production line, with a layer of yolk being added to the first ovum on day one.

On day two, another layer of yolk is added to the first ovum and a layer is added to the next ovum. This goes on down the line until the first ovum is mature and ovulation takes place. About 99 percent of the yolk is added in the seven to nine days before ovulation. At ovulation, the follicle surrounding the ovum ruptures and releases the yolk into the oviduct and the 25-hour process of forming an egg begins. The oviduct, a canal where the rest of the egg is formed, is about 25 inches long and has five distinct areas where different parts of the egg formation take place: the infundibulum, magnum, isthmus, shell gland, and vagina.

The first stop for the yolk is the infundibulum, a 3- to 4-inch funnel, which catches or engulfs the released ovum. The yolk spends about 15 to 30 minutes at this point where the chalazae are added to keep the yolk in the proper place in the middle of the egg. If you have a rooster, this is where the egg is fertilized. Regardless of fertilization, egg formation continues.

The yolk then moves down the oviduct to the magnum, a 13-inch-long area where the egg white forms, and then to the isthmus, a 4-inch-long area, where it is covered by the inner and outer shell membranes. This process in these two areas takes around four hours.

HOW COLOR COMES IN

From here, the forming egg moves into the shell gland where it will spend approximately the next 20 hours with the eggshell being formed around the inner contents. All eggs start out white because they are formed of calcite, which is a crystallized form of calcium carbonate that's naturally white. If you have a white egg–laying chicken, then no pigment is added after the shell is formed. If you have a blue egg–laying chicken, the blue pigment, oocyanin, is added here and sinks through the entire shell—so the outside of the shell and the inside of the shell are blue. If you have a brown egg–laying chicken, the brown pigment, protoporphyrin, is applied fairly late in the shell formation, after the white base has

been laid. Because it's applied so late, the brown pigment does not penetrate through the shell, leaving the inside of a brown eggshell white. For a green egg–laying chicken, the process is a little more complicated. It starts with the blue pigment being applied, followed by brown pigment. Since the brown pigment is applied late in the process, it doesn't sink through the entire shell, but it does mix with the blue on the surface to create green. The darker the brown pigment, the darker the green.

In the vagina, the bloom, or protective covering, is added to the egg. As the egg travels through the oviduct, it travels small end first. In the vagina, the egg is turned and moved to the cloaca where it is laid large end first.

> ## Fun Fact
>
> Near the shell gland and the vagina is the sperm host gland. This gland's sole purpose is to promote reproduction of the species. When a rooster mates with a hen, his sperm is stored here. When an egg is laid, a small amount of sperm is squeezed out of this gland and then travels up to the infundibulum where it can fertilize the next mature ovum. Sperm can live in the host gland for about two weeks. Hens can actually choose whether to keep sperm or not. Hens can "dump" sperm from a lesser rooster in favor of that from a lead rooster.

The cloaca in both hens and roosters is where the digestive, excretory and reproductive tracts all come together. So that a hen cannot lay an egg and defecate at the same time, the shell gland

Star Egg Carriers have dovetail construction with a metal rod handle. They're printed on three sides with one side left blank. That side is normally filled in with the proprietor's name.

Chicken History

For many backyard chicken keepers, it's all about the eggs: those nutrition-packed gifts from our hens. Nowadays, if you want to move those eggs from place to place, you stuff them into a cardboard or plastic egg carton for quick and easy transport to your family, friends, and customers. But it wasn't always like that.

In the early 1900s, eggs were just as sought after as they are today, and folks needed some way to safely transport their eggs. In 1903, a patent was issued for a wooden egg carrier with a removable cardboard insert and a metal slider that could join multiple stacks of carriers. These

Early egg cartons proudly proclaimed the chicken feed being used.

Early metal egg shipping crates came with an opening in the lid for the receiver's address. Inside was a sheet of paper that was turned over to reveal the address of the egg receiver or the address of the egg shipper.

Small nests of metal and cardboard, along with tissue paper, cradled individual eggs to keep them safe during shipping.

carriers were made by the manufacturing company of John G. Elbs with the STAR trademark. An advertisement from 1909 for the Star Egg Carrier shows that by then these carriers were really catching on:

"Savages carry eggs home in their hands or mouth. The Farmer's Wife makes the round of the mow, mangers, and hen house and carries the eggs home in her apron. That's no reason why you should use the old fashioned Paper Bag or Pasteboard box. Wake up! Join the procession. Star Egg Carriers mark the progressive store."

Star Egg Carriers solved the problem for local trips to buy eggs, but what if you didn't live close? In Fredericksburg, Virginia, in 1913, ordinances were passed that banned livestock, including backyard chickens, in the city for public health reasons. At the same time, similar ordinances were being passed throughout the country. People still needed fresh eggs and farmers had an abundance of them.

Inventor Stuart Ellis came up with the solution to connect farmers and city folk. He created a metal box that contained rows of cardboard bent into the shape of an egg and supported at the top and bottom with metal edging. Eggs were placed large end down, with tissue paper under and above.

The top of the metal carton had a rectangular cut out for the receiver's address. Inside were elaborate instructions on how to pack the eggs. These instructions were double-sided and had a place to write the receiver's address on each side. The egg buyer could send the empty carton back to the farmer with the flip of a sheet and vice versa. Pretty efficient!

There are all kinds of advertisements in seed magazines and early poultry magazines in the 1920s selling these crates, which held up to six dozen eggs and started at 85 cents each. They took advantage of the Parcel Post that allowed people to send crates, and not just letters, directly to each other.

Breeds and Egg Color

An egg collecting basket full of different colored eggs is achievable by picking the right breeds. There are lots of breeds to choose; many that are perfect for backyard chicken keeping and especially for families with kids. Below is a list of commonly found chicken breeds organized by egg color.

While technically not a true breed, Easter Eggers (top) are a popular choice as green egg layers. Australorps (bottom) are excellent brown egg layers. In fact, an Australorp holds the record for egg laying—364 eggs in 365 days.

COLORED EGGS

- Ameraucana (blue/green)
- Araucana (blue)
- Cream Legbar (blue)
- Easter Egger (green/blue)
- Olive Egger (dark olive to light teal)

BROWN EGGS

- Australorp
- Brahma
- Cochin
- Delaware
- Dominique
- Faverolle (light brown)
- Jersey Giant
- Naked Neck Turkens
- Marans (dark brown)
- New Hampshire
- Orpington
- Penedesenca (dark brown)
- Plymouth Rock
- Rhode Island Red
- Sussex (cream/ light brown)
- Wyandotte

WHITE EGGS

- Andalusian
- Ancona
- Appenzeller Spitzhauben
- Campine
- Catalana
- Fayoumi (tinted)
- Hamburg
- Lakenvelder
- Leghorn
- Minorca
- Polish
- Sicilian Buttercup
- Sultan
- Sumatra

stays wrapped around the egg while it moves through the cloaca. This shuts off the intestinal opening as the egg is laid.

Once you're familiar with how an egg is formed, simple math outlines a hen's laying cycle. After a hen lays an egg she will ovulate in the next half hour or so. Since the entire egg laying process takes 25 hours then she will lay an hour or so later the next day and so on. Eventually a hen will skip a day or two from egg laying to catch up. This composes a laying cycle and can vary from 12 days to almost a year depending on the breed of hen. Today's commercial hens are bred to have very little time between egg laying and ovulation, thus speeding up the process and producing more eggs during their laying cycle.

The Parts of an Egg from Outside to Inside

If you're lucky and you observe an egg just after it's been laid, you may notice it looks wet. That's because as a hen lays an egg, the very last addition is the bloom. This is a protective layer that starts out wet but when it's dry it protects the egg from dust, dirt, and bacteria. The bloom is critical to keeping the contents of an egg safe because the shell, while hard, is actually semi-permeable and allows air and moisture to pass through its more than 17,000 pores.

The shell is 9 to 12 percent of the total weight of an egg and is made primarily of calcium carbonate along with magnesium carbonate, calcium phosphate, and other organic matter.

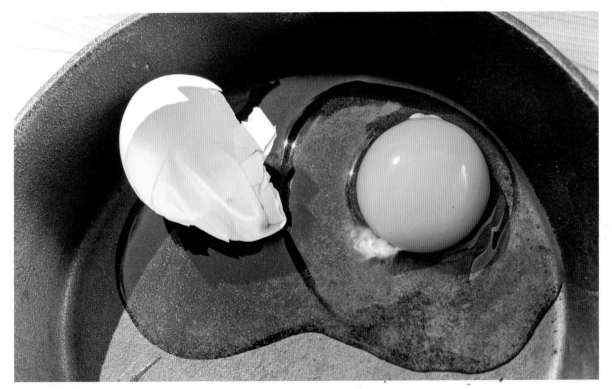

The different parts of this egg are clearly visible—the shell, outer membrane, egg white, yolk, and chalazae.

Once inside the shell, you will find inner and outer membranes, which are another line of defense against bacteria. These membranes are made of keratin and separate as the just-laid egg cools to form an air sac, usually at the large end of the egg. When peeling a hard-boiled egg, you can usually find this spot. When the membranes separate, the outer layer attaches to the shell and the inner layer surrounds the egg white or albumen.

Next comes the egg white, which makes up about 60 percent of the egg's weight. If you break a fresh egg in the frying pan, you'll be able to see that the egg white comes in two parts: the thin outer layer and thicker inner layer. The thick layer in a fresh egg is cloudy. This layer is a major source of riboflavin and protein. Over time it thins and becomes indistinguishable from the thin layer.

Also inside the whites of an egg are two chalazae. These are twisted cords that anchor the yolk in the center of the egg; the fresher the egg, the more prominent the chalazae. The cords are edible, although some people do remove them when cooking dishes that need to be perfectly smooth.

Last but not least is the yolk, the yellow portion of the egg. The yolk is held together by the vitelline membrane. The yolk accounts for at least 33 percent of the egg's weight and is packed with nutrition: proteins, vitamins, and minerals including iron, vitamins A, B, and D, phosphorus, calcium, thiamine, choline, lutein, folate, and riboflavin.

The color of an egg yolk is not an indicator of the nutrition value of an egg. Color is determined by what a hen eats. If she eats more foods higher in xanthophyll, a naturally occurring yellow pigment, then her egg yolks will be richer in color, sometimes almost orange. Many commercial foods contain marigold petals, which are high in xanthophyll. Other xanthophyll-rich foods include green leafy plants, corn, basil, pumpkins, carrots, peaches, prunes, and squash. The reason many backyard chicken owners find their egg yolks richer in color is because their chickens are more apt to free range and consume a varied diet that's high in xanthophyll.

Fun Fact

If you look closely at the surface of the yolk of a raw egg, you may see the germinal disc. This is a small spot, only 2 to 3 millimeters, where the sperm enters the egg. This is where the embryo develops and sends out blood vessels into the yolk for nutrition. The germinal disc is a solid white spot and is hard to see in an unfertilized egg, but in a fertilized egg it is referred to as a bull's-eye and is more prominent since it's a spot with a white outline and a clear center.

Collecting and Storing Eggs

There is actually quite a bit of controversy that surrounds egg collecting and storage. This is partly because so many people are used to seeing perfectly clean eggs in perfect egg cartons at the perfect temperature at the store. They have never seen an egg that's dirty or felt a warm egg that's just been laid. So knowing what to do with backyard eggs can seem more foreign than actually raising chickens. After all, we expect to have to study how to raise the birds, not how to handle something we see in the store every day.

The first step to egg collecting and storage actually takes place before the eggs are even laid: make sure your coop is clean. The cleaner you keep your nest boxes, the cleaner your eggs. It's a good idea to check your nest box bedding each day and clean as necessary. Also, be sure to keep a nice amount of clean bedding in your coop. This will help your chicken's feet get clean if they walk in from outside where the ground may be wet and muddy.

But even with the best of intentions, things happen and eggs can and do get dirty. It's important not to wash an egg because that will remove the bloom and leave the outside of the eggshell

unprotected. If you have some small spots of dirt, just use a dry cloth and wipe it off. If the shell is too dirty to get clean that way, you can rinse it in warm, not cold, water. Then refrigerate that egg and try to use it quickly.

Tip

For really dirty eggs, I don't like to store them, and I don't like to wash them. So, I feed them to my dog and even to my chickens. I usually cook them immediately and feed them scrambled. They are a great source of nutrition for my animals, and they leave my dog's coat shiny and healthy.

Once your eggs are collected, it's an individual decision whether to refrigerate them or not. In Europe and Great Britain, eggs are not refrigerated. This makes sense when you are familiar with a hen's broody behavior. Think about it, a broody hen can take over a week to lay her clutch and her eggs remain viable during that time without refrigeration. In America, it's generally accepted that eggs should be refrigerated. When you leave them at room temperature, eggs will start to go bad faster than eggs that are refrigerated. So, if you're going to use your eggs quickly, then it's perfectly safe to leave them out. If not, refrigeration is your friend. Just remember that if you choose refrigeration, then you need to stick with it. If refrigerated eggs are left at room temperature for long, they will sweat and this increases the movement of bacteria into the egg. It's recommended not to leave refrigerated eggs in room temperature air for more than two hours.

Egg cartons are available in different sizes to hold your backyard eggs. Just remember to store the eggs large end up, pointy end down.

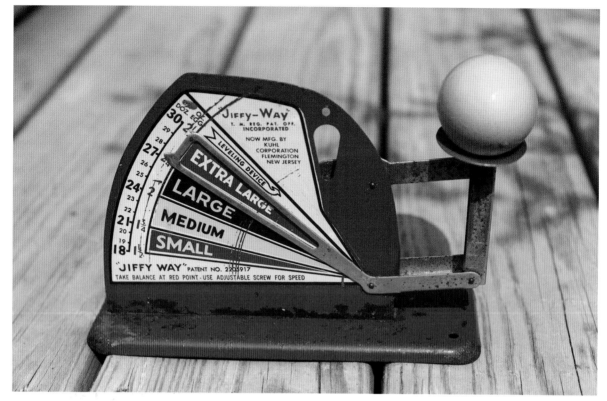

An antique Jiffy-Way scale identifies this egg as extra large.

Fun Fact

Hard-boiled eggs spoil faster than uncooked eggs because boiling removes the bloom and leaves the pores in the eggshell open for contamination. Make sure to refrigerate hard-boiled eggs within two hours after you've cooked them and use them within a week.

Egg Size and Grading

My Jiffy-Way egg scale is one of the most fun items I have in my vintage chicken-keeping collection. Jiffy-Way spring scales were originally made in Owatonna, Minnesota, starting in the 1940s. Mine is a Jiffy-Way scale but it says, "Now MFG. by Kuhl Corporation Flemington New Jersey" so it's a later version. Mine is also missing a few pieces as many of these older scales are, but it still works.

Egg scales didn't gain popularity until World Wars I and II. Until then, egg weighing was done as a way to select eggs for hatching. But during the wars, the U.S. War Department bought eggs to feed the troops and paid for them based on their size. Egg weight equaled cash and created the need for scales. During that time, egg scale advances and production were at a high.

Today you can still buy Jiffy-Way and other egg scales as they're an important part of serious baking and egg-selling efforts.

So let's look at exactly what is egg sizing versus what is egg grading. Although grading is often used generically as a term for classifying eggs, technically they are not one and the same. Egg size and grading standards are set by the United States Department of Agriculture. Egg sizing determines where an egg fits in six different weight categories; jumbo, extra large, large, medium, small, and peewee. (My scale is older so it only has four categories.) As a point of reference, most baking recipes are based on large eggs. Note that size is determined by the average weight in ounces per dozen. That's why if you look hard enough, not all eggs in a carton at the store are exactly the same. You can see small differences.

EGG SIZES

Size	Ounces per dozen	Ounces per egg
Jumbo	30	2.50 or more
Extra Large	27	2.25 to 2.50
Large	24	2.00 to 2.25
Medium	21	1.75 to 2.00
Small	18	1.50 to 1.75
Peewee	15	1.50 or less

Egg grading actually has nothing to do with weight or the nutritional value of the egg. It is a measurement of the quality of the outside shell and the inside contents. The grades are in descending order from AA, A, and B. It is mandatory that eggs are inspected for wholesomeness, but grading for quality is voluntary. So if a company wants to have its eggs graded, they can pay the USDA for this service and a USDA grade shield is on the carton. If companies do not use the USDA, then compliance is monitored at the State level and the carton is marked with a grade but not the USDA shield.

Grade AA eggs: High and round yolks, thick and firm egg whites, perfectly shaped shell that is free from defects.

Grade A eggs: Almost the same as AA eggs, but the whites are considered reasonably firm.

Grade B eggs: Yolks are wider and flatter, whites are thinner. Shells may have slight stains. Not often found in the store. These eggs are usually put into other products.

Testing Egg Freshness

At some point we've all put a carton of eggs in the refrigerator or on the counter and lost track of how long it's been there. Many culinary experts will tell you to float your eggs in cold tap water to see if they are fresh or not. While this isn't a foolproof method, there is science behind it.

As an egg ages, water vapor and gasses are released through the porous shell. This is what causes the yolk and whites to shrink and makes the air sac larger. Some air does enter the shell as water vapor and decomposition gasses leave, but the overall mass of the egg is reduced.

So a fresh egg is heavier and it will sink to the bottom. As the egg gets older it will start to stand up in the water, usually with the large end at the top since that's where the air sac is located. The

An older egg will float in water while a fresh egg will sink to the bottom. This doesn't indicate whether an egg is good or bad, just how fresh it is. It's a great test to use if you forget to label your eggs when they're collected!

oldest of eggs will rise completely off the bottom and float in the water.

This only means that an egg is old, but does not mean it's spoiled. That's why it's always important to crack your eggs in a separate bowl. If your egg has an unusual odor and appearance, then it's not safe to use.

Laying Timelines

So now that you know how an egg is formed and how to store it, when can you expect to see eggs from your hens? This will vary from breed to breed but in general a hen will start to lay around 18 to 21 weeks of age and she will reach her best production around eight months of age. Heritage breeds can take a little longer and generally begin egg production around 24 weeks old.

Egg production is based on a bird's maturity plus the amount of daylight available. Spring chicks will mature and begin laying in the summer. Fall chicks, however, won't reach maturity until the winter when daylight is in short supply. Even though they are mature, they won't lay in the winter. Typically, they will begin laying in the spring once daylight increases.

Q&A: All About Eggs

How should my eggs be stored in the carton?

It's important to store your eggs with the large end up, pointy end down. By doing this, your egg will stay fresher longer because less moisture will be lost since the air sac is located on the large end. If you're storing hatching eggs, they must be stored with the large end up so the air sac stays in place.

How long do eggs stay good in the refrigerator?

Properly stored and refrigerated eggs rarely spoil. Over time, they will actually dry up. But with that said, a rule of thumb is that eggs can stay good up to 100 days in the refrigerator.

If I add a rooster to my flock, how long will it take for me to get fertilized eggs?

Technically it takes only two days before a hen will start laying fertilized eggs after being introduced to a rooster. Realistically, it may take your rooster some time to mate with each hen so a good rule of thumb is about a week after adding a rooster, you'll have fertilized eggs.

If I have more than one rooster, how do I know which eggs come from which rooster?

The only way to be sure is to separate your birds so you know who mated with whom. Interestingly, in a multiple-rooster flock, hens can decide which roosters fertilize her eggs. Hens play a subtle, but influential, role in the genetic destiny of the roosters because they have the ability to dump sperm. Hens don't always have a choice about whom they mate with. Hens and roosters are not mutually exclusive by forming pairs. And it can be hard for a hen to fight off unwelcome advances. So, a hen accepts this fate but can choose to dump the sperm of a lower-ranked rooster and keep the sperm from a more dominant rooster. This is Darwin's theory in practice and ensures reproduction of the fittest.

Chicken (and Egg) Health

I usually go out to the coop a few times throughout the day to check on my flock. One afternoon I unexpectedly found Little Muff, one of my Easter Eggers, on the floor of the coop by herself while everyone else was out roaming the yard. (Chickens lying on the coop floor in the middle of the day are usually not a good sign.) My heart skipped a beat and I looked closer. She was alive but the entire top of her head was bleeding. At first I thought I could see her brains!

I quickly gathered her up and brought her inside our mud room, the spot for most chicken examinations. I keep a rubber container handy that's just the perfect size for a chicken. That day I filled it with a layer of bedding since I knew she was in shock and needed the warmth. It was early spring and still quite chilly.

As I inspected Little Muff more closely, I realized it wasn't her brains that I could see, but it was pretty close; portions of her skull were exposed. Her head had been mercilessly pecked by someone in the flock. My theory is that our meanest rooster, who is now living in a bachelor rooster flock, tried to mate with Little Muff and accidentally caused some bleeding. Once that happened, he went into a frenzy and started pecking her.

Luckily I had the knowledge to carefully treat Little Muff's wounds. I washed the blood away from her eyes and feathers. I used warm water to gently loosen the caked areas, being careful to leave the original scabbing. It took a while because I had to avoid her eyes and nose, but once she was clean, I used a cotton swab to apply Neosporin. I set Little Muff up in a crate in the mud room with bedding, water, and food. She settled in and seemed to enjoy to the time to recuperate. Each day I added Neosporin to the healing wound to keep it moist. Once Little Muff was on the road to recovery, she moved to the garage and I brought our other Easter Egger, Big Muff, inside to keep her company. Eventually Little Muff and Big Muff rejoined the flock, minus our mean rooster. And, after molt in the fall Little Muff grew enough feathers back to cover her once-bare head.

Wound care is discussed in further detail on page 101, but the point of this story is that once you have a flock you will likely want to learn how to take care of small emergencies on your own. In this chapter, we'll cover some of the basics as well as chicken and egg warning signs. I'll also cover many of the day-to-day things you can do to maintain a healthy flock. You'll find that just like human health care, an ounce of prevention is often the best medicine.

Daily Check-Ups

As opposed to cats, dogs, and even larger livestock, veterinarians who have expertise with chickens and other poultry can be hard to find. So while in many cases a veterinarian is your best choice to handle a particular problem, you may not have that luxury. At some point, you'll probably need to deal with an emergency. You'll certainly have to handle general health issues and routine preventative care.

No matter what, it's essential you know your birds well. Daily check-ups fall under preventative

It's a good idea to give your chickens a quick once-over each day to make sure they're doing well.

care and will help you to know what your birds look and act like when they're healthy.

This is important because when we humans are sick, we'll usually tell someone. We might head to the doctor and try to rest and recuperate as our schedule allows. But we're a little different than chickens. Chickens are prey animals and flock animals. To show weakness makes them vulnerable to predators and can knock down their place in the pecking order. This means chickens will hide an illness as long as they can.

The problem with this is that if you don't pay very close attention, you often won't notice a chicken is sick until it's way past the point of being saved. That's why a daily once-over just to see how things are going makes a lot of sense. It doesn't have to be complicated, but it does involve observation. This is something I've noticed many of us have forgotten how to do. We're so busy multi-tasking that we may just open the chicken door in the morning, drop off some fresh food and water, clean the nest boxes and come back and do it all again at night—all the while never really noticing what's happening with our chickens.

There are telltale signs that your chickens may be sick or something is off with your flock.

You may see one of these signs or a combination of a few. Here's a list of signs to watch for:

- Consistently laying odd eggs in the nest boxes

- No eggs

- Chickens that are droopy and lethargic

- Pale and droopy combs and wattles

- Lack of appetite

- Excessive thirst or no thirst at all

- Consistent diarrhea or bloody droppings

- Feces that are coating feathers on your chicken's back side

- Missing feathers

- Lack of ability to roost

- Limping

- Bloody cuts or spots

- Loss of breast muscle

If you spot a chicken that's exhibiting signs of illness or injury, it's important not to panic. You may want to isolate that chicken and further investigate the problem. Be smart and be deliberate. Chickens are hardy, but they can get sick. It is distressing to have a sick chicken—not all problems can be cured or solved. Luckily, many can. It's good to have trusted resources on hand to consult if you're unsure. It's also a good idea to have your veterinarian's phone number easily accessible. If you have no veterinarian, your local extension agency can be an invaluable source of information.

Troubleshooting Egg Production

Eggs are the goal of a backyard chicken keeper and when you don't get them, it's frustrating. Yet the reason why you're lacking eggs isn't always cut and dry. It can take a little patience and some detective work to figure out what's causing a dearth of eggs. Of these many reasons, some are more obvious than others. Here's a list of possible causes:

Broodiness: A hen does not need a rooster around to go broody. Broodiness is inherent to some breeds; they instinctively want to hatch and raise chicks. When a chicken goes broody, her pituitary gland releases a hormone called prolactin, which stops the hen from laying eggs. If egg production is important, a broody hen can present a problem for the chicken keeper. Broodiness can also be hard on a hen's body since they eat and drink very little during that time. They are also more susceptible to mites and lice as they are in one nest all day, every day. Many chicken keepers will try to break a hen of her broodiness. Sometimes easy techniques—such as not letting eggs accumulate in the nest and repeatedly removing the broody from the nest—work quickly and well. Another technique that requires a little more commitment from the chicken keeper is to remove the broody and put her in a wire-bottomed, but safe, cage. In theory, this will cool down her breast, which is warmer when she is broody, and stop her broodiness after a few days. While admirable, these techniques

Eggs are a good source of protein and the shells are a good source of calcium. Don't be afraid to feed them to your chickens. However, make sure your chickens have free-range time and stimulation so egg eating doesn't become a fun pastime.

don't always work, so sometimes it's just better to let nature take its course and let the hen remain broody until she's done. Just make sure to provide her with a clean nest box and adequate food and water so she can stay healthy. And remember, the longer a hen stays broody, the longer it takes her to get back into an egg-laying cycle.

Tip

Broody hens don't always lay their clutch in your coop's nest boxes. Broody hens are infamous for laying a clutch in hidden places. So keep a watchful eye out. Hens have been known to disappear and then come back over three weeks later with a new brood in tow.

Egg Eating: Egg eating is a nasty habit that's difficult, if not impossible, to break once it starts. There are two main reasons for egg eating: protein deficiency and lack of space. This is a behavior that starts innocently enough when an egg accidentally breaks; the contents spill out and it's too tempting to resist. The hen that discovers eggs are a tasty treat shows everyone else and soon your whole flock is eating their own eggs.

The best cure for egg eating is to prevent it in the first place. Feed your birds a well-balanced layer ration and give your birds room to roam. If they can't free range, give them access to the outside with a large enclosed run. Make sure your hens are laying hard-shelled eggs. Collect your eggs often and line your nest boxes with lots of clean cushioning using straw and wood shavings. Keep your nest boxes darkened and add calming herbs, such as lavender, to make sure the hens lay eggs in peace. Some people try to show the hens that eggs aren't tasty, so they add golf balls and wooden eggs to the nests. If all else fails, there are nest boxes that tip down in the back so the eggs naturally roll away from the nest box into a safe collecting area leaving nothing for the hen to eat.

Nutrition: Chickens have to have adequate food and water to keep their egg production at its peak. Dehydration, even for short periods of time, can affect a hen's laying rate. Make sure clean water is always in good supply since birds like to drink small amounts often. Always make sure to offer a properly balanced food free choice along with proper calcium supplements and treats. Make sure that enough food and water stations are offered so even the lowest chicken in the pecking order has easy access.

Time of Year: A hen's laying hormone is triggered by daylight. They need 14 to 16 hours of daylight each day to maintain maximum production. In the fall and winter, daylight is in short supply so egg production will decline or stop altogether. In summer's hot weather, chickens may also stop laying eggs.

Predation: When you open the coop in the morning it's exposed to all kinds of predators. Don't always assume the egg thefts are caused by wild predators. Backyard dogs are infamous for sneaking into the coop and grabbing a snack. On a personal note, our dog Sophie was caught sneaking into the coop through the pop door. Her personal high was eating 14 eggs at one sitting. Needless to say, a smaller pop door solved that problem.

Molting: As hens molt, their reproductive system takes a rest and they stop laying eggs since their bodies are hard at work making new feathers. (More on molting can be found on page 155.)

Stress: Many things such as predation, poor nutrition, poor health, and overheating can cause a chicken physical stress and result in fewer eggs. But it's the mental stress that people often overlook. Chickens are creatures of habit and anything out of the ordinary can cause anxiety. Did you put a new light in your yard? Is a new car or truck going by on a nearby road that makes lots of noise? Did the neighbors get a new dog that barks at odd hours? For each backyard, the questions will be different. But it's important to think like a chicken and you may discover something you thought was small is causing a big problem.

Health: Checking egg production is an important part of a chicken keeper's daily health checkup. Viruses, bacterial infections, mites and lice can all stop a hen from production. A lack of eggs can be a symptom.

With multiple hens in a flock, it can be hard to determine who is laying eggs and who's not since you're not likely to see your hens in the act of laying an egg. If you've got different colored egg layers, you'll be better able to tell. Say you have two green egg layers in a flock of brown egg layers, if you find no green eggs routinely or just one green egg routinely then you can target a problem with your green egg layers. There are physical clues to help tell whether a bird is laying or not. The easiest clue is to look at the vent. A chicken that's not laying will have a small, tight, and dry vent. A chicken that's laying will have a large vent that's moist and oval shaped. On a laying hen you should also be able to fit three or more fingers (two for a small breed) in the area between the pubic bones. This indicates space for the flexibility needed when laying eggs.

Imperfect Eggs and Oddities

Most of us are used to buying eggs from the grocery store before we buy our first chickens. At the store, all the eggs are uniformly sized, perfectly colored, and have no oddities. Little did we know this is a distorted reality and not the way chickens really lay eggs!

When you start raising backyard chickens, you quickly realize that not every egg looks like a perfect grocery store egg. This doesn't mean an imperfect egg is necessarily bad; in fact, imperfect eggs are used in the commercial egg industry too. Those eggs just often end up in places where they can't be seen, such as cake and cookie mixes. In a backyard flock, imperfect eggs often require a picture being taken to mark the occasion and maybe even a dinner conversation.

However, while they're fascinating, it's good to know what they signal as well. The fact of the matter is that most odd eggs are just that; odd, but perfectly fine to eat. If you consider the math, a hen can lay over a thousand or more eggs in her productive years and it's almost impossible to achieve grocery store perfection over time and in that quantity. We shouldn't expect a hen to lay

You can get a good sense of size when this double-yolk egg and extra small eggs are sitting next to a penny.

A double-yolk egg when cracked open.

a perfect egg every time she lays an egg. On the other hand, if you get odd eggs on a regular basis from one hen or all at once from many flock members, then it may be a clue that something is wrong.

Double-Yolk Eggs (or Multiple Yolks): As spring chicks start to grow up and lay eggs, you may eventually find a double yolk egg in your egg-collecting basket. To be clear, double yolks are just that. When you crack a single-shelled egg open, you'll find two yolks inside.

Double-yolk eggs, or in rare cases, triple-yolk eggs, are formed when a hen releases two or more yolks at once. They move through the reproductive tract together and are encased by a single shell.

Many times you'll know if your hen laid a double yolk egg because it will be much bigger than a standard-size egg—double or even triple the size. Ouch!

Double-yolk eggs can happen as chickens begin their laying cycle. At that point, a hen's body is trying to get into a normal rhythm and misfires can happen. Double-yolk eggs can also happen at the end of a hen's laying cycle as things are winding down, causing her normal body rhythm to be thrown off beat.

Although double-yolk eggs may have a softer shell, they are not truly soft-shelled eggs. The same amount of calcium goes into a double-yolk eggshell as goes into a standard-size shell. That may cause the shell to be a little softer than normal.

Statistically speaking, double-yolk eggs happen in about one in every thousand eggs. They are more common in hybrid and large breed chickens and can be hereditary. (In some countries, hens that lay double-yolk eggs are prized and bred to lay them.) Although they are not a sign of illness, double-yolk eggs can cause problems for the laying hen. Consistent laying of extra-large eggs can lead to egg binding, where the chicken cannot pass the egg and it's stuck inside or causes a prolapse, where part of the oviduct is sticking out of the chicken's vent on the outside of her body.

Extra-Small Eggs: These eggs go by many names including fairy eggs, fart eggs, rooster eggs, and wind eggs. If you crack one open, usually you'll find there's no yolk inside, although sometimes they'll contain a yolk. These eggs are normally laid by young hens whose reproductive cycle is trying to get in sync or by older hens whose reproductive cycle has been disturbed. When this happens, a piece of tissue accidentally enters the reproductive system and goes through the egg making process.

Soft or Shell-less Eggs: Warm weather brings thin shells as chickens pant. That's why you see more soft or shell-less eggs in the summer. Panting helps cool the chicken but causes a reduction in calcium being put into egg production. A soft shell, or an egg with just a membrane, can happen when a hen rushes laying an egg. They occur in young hens as their bodies are adjusting to laying. Soft shells can be a signal that your hens need more calcium or Vitamin D. They can happen with chickens at peak production as their bodies are working hard to keep up with the demands as well as with older chickens. If they are laid regularly, they can be a sign of disease like Newcastle disease or infectious bronchitis. Wrinkled eggs can be a sign of infectious bronchitis or Newcastle disease. Soft eggs can signal mycoplasma. Shell-less eggs can signal avian influenza.

Rough Eggs: Pimples, bumps, and rough patches on eggs are caused by calcium deposits. They can

This egg has a cluster of calcium deposits at its pointy end.

signal your hen is getting too much calcium. This is more likely in winter as your hens are eating layer feed with increased calcium but not laying eggs because of the season. Rough eggs can also be a sign of stress caused by lack of water or environmental disturbances like loud noises. These stresses can cause eggs to stall as they go through the egg-formation process. When they're stalled, extra shell can be added. Chickens also need vitamin D to process calcium correctly. Sunlight is the best source of vitamin D, so make sure your hens have access to the sun.

Wrinkled and Cracked Eggs: Sometimes more than one egg will move through a hen's reproductive tract; the egg behind can bump into the first egg, causing damage in the form of wrinkles and even cracking that gets repaired before the egg is laid. Also, as an egg moves through the reproductive process, it can be damaged through rough handling that puts pressure on the forming egg. The hen's body will "check" these eggs and try to repair the damage, but they usually are laid looking rough. Wrinkled eggs can be a sign of infectious bronchitis or Newcastle disease.

Egg within an Egg: This is a rare phenomenon and the cause is unknown. But an egg that's about to be laid is moved back up into the reproductive tract where more egg white, possibly another yolk, and a shell is added around it. This is an oddity most chicken keepers will not ever see. If you do, it's definitely picture-worthy!

Eggshells with Spots: As color is being applied to a forming egg, the egg can rotate too quickly or too slowly. If it rotates too slowly, the egg can have spots or speckles. If it rotates too quickly, the egg can have smears of ink.

Meat Spots and Blood Inside Egg: These abnormalities are not normally seen in grocery store eggs because they are candled before they are packaged. Candling is not 100 percent accurate, as darker eggshell colors make it harder to see inside the egg. But it's effective enough that most modern consumers have never seen these abnormalities. Eggs with meat spots and blood spots are technically fine to eat. A meat spot is caused by a small bit of lining being released as the egg is being produced. A blood spot occurs when a blood vessel ruptures and a bit of blood is released along with the yolk. You can pick out these spots before eating or choose not to eat the egg and instead scramble it up for your pets and/or flock.

Understanding a Lash Egg and Salpingitis

If you've never heard of a lash egg, chances are you're in good company. Lash eggs can be a one-time occurrence or an uncommon symptom of an illness that is actually the number one killer of laying hens—salpingitis, an inflammation and infection of the oviduct.

Although known as a lash egg and having the appearance of an egg, a lash egg really isn't an egg at all. These masses are produced when a hen sheds part of the lining of her oviduct along with pus and other materials. Lash eggs travel through the reproductive system, so they are often egg shaped. But when they're cut open they reveal layers of tissue and other material. They have been described as having a rubbery feel, with the insides being the consistency of cooked yolks and having a cheesy look. Each lash egg is similar but different. Some are a tan to yellow color, while others can even be greenish.

Salpingitis is caused by a bacterial infection that travels to the oviduct. Once illness sets in, salpingitis, or inflammation of the oviduct, occurs. When a hen has an inflamed oviduct, her body may try to contain it by forming a lash egg and trying to expel the infection.

Salpingitis is not always a death sentence for your hen. Sometimes hens have a strong-enough immune system to beat the infection and inflammation on their own. It can be a one-time occurrence. For some, if caught in time, they can recover with the help of antibiotics. The symptoms of illness are vague, including lethargy, laying multiple soft eggs, a droopy off-color comb, and excessive thirst. Hens with salpingitis will sometimes walk like a penguin since their swollen abdomen prevents easy movement. A lash egg is also a symptom.

If you're unsure of what's happening with your chicken, it's a good idea to take her to the vet where you can be advised on the best course of action.

Most often by the time this disease is caught, it's pretty far along. Many hens don't live much longer than four to six months as the infection festers, the lash egg grows inside them, and their organs are pressed too much to continue functioning. That's why you may see a chicken having a hard time breathing as this condition progresses.

If a hen recovers, her productivity can be compromised. She may never lay again or will lay far fewer eggs in the future. For a backyard flock, this is normally not a problem as fresh eggs are a benefit of having chickens but not a requirement—many flock members take on pet status. On the other hand, in a commercial chicken operation, a chicken that lays a lash egg is culled. When egg production is the goal and makes your bottom line, a reduction or stoppage in laying can't be tolerated.

There is no surefire way to prevent salpingitis in your hens. It is most common in birds at two to three years old. Make sure your chickens are getting a healthy diet and free range exercise each day. Good animal husbandry practices, such as keeping a clean coop and providing clean water and fresh food, can be helpful in preventing the spread of bacteria. But even with the best chicken keepers, this condition can still occur.

Chicken Droppings

If you've got chickens, you've got poop. While it may not be pleasant to talk about, chicken droppings can give us many clues about our birds' health. It's not something to obsess about, but it is something to watch. That way you'll know if something is abnormal.

Chicken droppings can actually be confusing as they often look different throughout the day and throughout the year. This is normal. Chickens also have two types of droppings, intestinal and cecal, so let's start there.

Intestinal droppings are made of feces and white urine salts. Chickens don't urinate as

A healthy intestinal dropping made of feces and white urine salts.

humans do, so both of their excretory functions are combined into one. Passing urine salts, or urates, eliminates nitrogen and water from the body. This is actually an adaptation for flight, so that a bird can remain as light a possible by not storing excess water as liquid urine. The feces portion of an intestinal dropping varies in color depending on what the birds have been eating. Typically it's brown to green. Intestinal droppings are excreted during the day and night.

Cecal droppings come from the ceca, which are two pouches located at the crossroad of the large and small intestines. The ceca are essentially a fermentation vat where moisture is absorbed from digesting food and that food is broken down by bacteria that live in the ceca. Cecal droppings happen less than intestinal droppings. They are smelly and can be different colors from yellow to black and brown. They have a different texture than intestinal droppings, which are more firm.

Cecal droppings are smoother and shinier and can be described as a little runnier. It's important not to confuse cecal droppings with diarrhea.

Even within two main types of droppings, chicken poop can have many variations. It may not be tasteful to bring up, but it's the same as in humans. What you eat influences the appearance of your stool and it's the same with chickens. For instance, free-range chickens that have access to vegetation often have droppings with more of a greenish appearance. If your chickens have been munching on watermelons, their droppings may have a more reddish appearance. Many people watch for blood in their chicken's droppings. But don't get confused between blood and a little bit of red in the feces. A little red is just fine and is the normal shedding of intestinal lining. Sometimes chickens can even pass clear liquid with no actual feces in it, just a few white urates hanging around. If you have a more extreme or unusual change, look around for what might be the contributing factor. For example, I have a friend whose chickens' droppings were black in color. It turns out they were consuming the wood ash from their dust baths over the winter!

Consistency can also vary. For example, a broody chicken will often hold her waste in her body and defecate only once or twice a day. This is called a broody poop and it's huge and smelly! On the other hand, on hot days chicken droppings can be much more runny and watery since they're consuming more water to stay cool.

The moral of the story with chicken poop is not to become overly obsessed by your chicken's droppings. But do take note and watch. By knowing your chickens, you'll know when changes in droppings are occurring. You'll be better able to tell whether diet or weather influenced these changes or whether they indicate a bigger problem.

Coccidiosis

Coccidiosis is a leading killer of baby chicks over three weeks old and is a common killer of adult chickens as well. So what exactly is coccidiosis? This illness is actually caused by a parasite known as coccidia. These parasites cannot be eliminated from your chickens; in fact, they are present in small numbers in every chicken. As chickens digest and excrete their food, the coccidia eggs, known as oocysts, are shed from the chicken's intestinal lining and passed out of the body. Outside the body, the oocysts mature and are often eaten by other chickens as they scratch and peck. The coccidia make their way to the intestines where they begin reproducing. So the cycle continues and coccidia are easily passed from chicken to chicken.

There are different types of coccidian, each of which inhabits a different area of a chicken's intestinal tract. Chickens can have more than one type of coccidia in their bodies at the same time. Chickens can actually become immune to the coccidia they are exposed to in their environment by either surviving a case of coccidiosis or through gradual exposure. The reason it kills baby chicks so quickly is that they have not had a chance to gradually build up immunity and their immune systems are not fully developed. Using a medicated chick starter for their first four months or immunizing your chicks gives their immune systems a chance to fully develop and keeps the number of oocysts in their bodies down. Adult birds can shed billions of coccidian oocysts, so the onslaught is hefty. At three to five weeks, young chicks are the most susceptible to coccidoisis. Older birds with compromised immune systems or under stress are also susceptible to coccidiosis.

Chicken droppings with large amounts of red and blood can be a *big* problem because they may indicate a case of coccidiosis that is about four to five days along. At this point, the disease is at its highest point and this is the time that most deaths happen. Before that, the signs of coccidiosis are a little vague with birds being pale and listless, often with ruffled feathers. They'll probably have diarrhea and you may see poor development in a young chick.

If your chickens contract coccidiosis, it's important to treat them quickly. There are a variety of drugs that can be used to treat coccidiosis. Make sure to follow the labeled directions carefully so you don't overdose your chickens.

Coccidia, while a normal part of the environment, thrive in hot, humid, and dirty conditions. It's important to keep your brooders clean and dry. It's essential to keep your waterers filled with fresh, clean water throughout the day as droppings in the water can spread the coccidia oocysts. And make sure your feeders have only fresh food without droppings in the food.

The key to developing immunity to coccidiosis is making sure your birds are healthy. Young chicks can be gradually exposed to the coccidia they will face in the backyard by adding small clumps of grass with some dirt attached right in your brooder. Let the chicks have fun pecking and scratching through the clumps.

Pasty Butt

Pasty butt is known by other names, such as vent gleet and pasting. You may have gotten through raising your first flock of chicks and never experienced this problem. But as you add to your flock, it's good to know about pasty butt so you can treat it.

Pasty butt is a problem with baby chicks that have experienced stress, such as temperature fluctuations in the brooder, over-handling, and chilling while in transit. That's why it's a much more common problem with chicks that have been shipped.

However you get your chicks, you can help prevent pasty butt with a few simple steps. Before your chicks first arrive, it's a good idea to have their water already set up and in the brooder. This will allow the water to adjust to room temperature and prevents them from drinking cold water and getting chilled. Then let your chicks drink water for the first 30 minutes or so before adding food to the brooder. This helps add fluid to the digestive system and keep droppings from being too dry.

How can you tell if your chick gets pasty butt? Well, pasty butt is just what it sounds like: a chick's vent area becomes pasted with feces. In mild cases, pasty butt is not a problem and is really more bothersome to the chicken keeper since it's messy. But in more severe cases, the vent becomes completely blocked and the baby cannot clear its excrement. This can lead to death if it's not treated. If your chick or chicks have pasty butt, you need to assess the situation. If there's just a little bit of mess, you can choose to leave the chick as it is and monitor it closely. If there is too much mess or a blockage, you must first remove the droppings from the chick's vent. This is where caution must be used. The droppings can really be stuck hard onto a chick's vent. If you pull off the stuck plug, you can tear the tender vent skin and do great harm to your chick. The best thing to do is to first try gently wiping with a moist cloth. If the droppings are still stuck, then you can hold a warm, wet cloth to the area to soften things up. I once heard a poultry speaker tell the story of coming home to find his wife gently holding a chick with pasty butt over a mug of warm water that had some steam. He thought this was a great idea for gently softening stuck droppings until he noticed she was using his favorite coffee mug. Yikes!

Tip

Don't get confused by the drying or dried umbilical cord from a chick's naval. This is located toward the belly of a chick, further down the legs. The vent is directly underneath the tail. If you see the remains of an umbilical cord, never remove it! Let it dry and fall off on its own.

Once the vent is clear of droppings, use a cotton swab to lubricate that area with some olive oil or petroleum jelly. This will help to keep the droppings from sticking in the future. Reapply as necessary and make sure to keep that area clear of droppings. Within a few days of treatment you should notice marked improvement in your chick's ability to excrete normally.

Worming

Worms are a tricky area for chicken keepers. They prompt lots of questions from confused chicken keepers and lots of anger as people's opinions differ on prevention and treatments.

Worms are a natural part of the outdoor world and chickens and worms can coexist in a happy medium. In fact, most chickens will carry some load of worms throughout their lives. When a chicken is stressed though, that balance can tip in favor of the worms and that's where problems occur.

Before we discuss treatments, let's talk about what worms are out there and how your chickens can get them. The worms most people talk about are intestinal worms, but it is important to mention there are also worms that invade other parts of a chicken's body. Eyeworm lives in warm climates and will burrow into a chicken's third eyelid and cause it to swell. Gapeworm burrows in the windpipe and causes chickens to gape or gasp for air as they breathe. Common worms can be broken into two groups, roundworms, which also include thorny-headed worms, and flatworms, which include tapeworms and flukes.

As with any parasite, it's important to know the life cycle of worms. Some have a direct life cycle, which means the adult worm lays eggs in the chicken, the eggs come out in the chicken's droppings, and other chickens eat those eggs. Some have an indirect lifestyle, which means the adult worm lays eggs in the chicken, the eggs come out in the chicken's droppings, and then other hosts such as insects eat those eggs

Squash can be a natural dewormer, and it's a healthy treat your chickens will love.

and the chicken eats those hosts. (Simply put, chickens get worms from things they eat off the ground.)

So what does a chicken with an out-of-balance worm load look like? Symptoms can include pale combs and wattles, lethargy, reduced egg laying, and weight loss. Sometimes chickens with excess worms can also have foamy diarrhea or green droppings.

So how can you prevent worms? Honestly, you'll never prevent all worms. But you can tip the scale in your chickens' favor by practicing good animal husbandry. Keep your coop and run clean and don't overcrowd your birds. Make sure your chickens have a healthy diet that includes vitamin A, which can make your chickens more resistant to a worm attack. You can try to eliminate intermediate hosts, but if your chickens free range that's nearly impossible. Still, do try to make sure your chickens have a large area to free range and, if possible, rotate that range. For instance, my chickens free range inside a fenced area but I do keep other areas just outside the fence mowed. If I'm going to be home for the day and can be extra vigilant, I let the chickens into these extra areas. Ideally I like to do this as many days as possible to give the normal chicken yard a nice break.

> **Tip**
>
> Try not to throw feed and treats on the ground for your birds to pick up. Put the treats in a container. This will decrease the time your chickens peck at the ground and possibly pick up worms.

There are many chicken keepers that swear by natural methods of worm control. I'm always a firm believer that natural methods should be tried as a first resort and as preventative maintenance. Seasonally, there are healthy and natural treats you can give your flock that combat worms. If you find they work well controlling worms for your flock, then so much the better. If not, your chickens are no worse for the wear. My favorite is to give my chickens pumpkin and squash (complete with seeds) in the fall. In fact, I buy extra pumpkins and store them as long as possible to stretch out this treat. In the spring and summer, I feed my chickens lots of watermelon and cucumbers. And, year-round, I give garlic to my chickens.

Many chicken keepers use food-grade diatomaceous earth (DE) as a worm preventative and swear by it. Diatomaceous earth is a white powder that's made by grinding the fossilized remains of algae. It works by preventing larvae from maturing into adults, thereby breaking the egg-laying cycle. The recommended ratio is one pound of DE to every 50 pounds of feed.

Worming twice a year using medications just because that's what farmers have always done is not a good strategy. In the long run, treating for worms twice a year and switching up medication can set up drug resistance. This is not good if you ever truly have a problem. If you do suspect a chicken is suffering from an overblown infestation, then it's a good idea to take a fecal sample to your veterinarian. Have the vet diagnose which worm is causing the problem. The vet can prescribe medicine that treats the specific type of worm that is causing the problem. While worming medications generally try to be broad-spectrum, if you treat for the wrong worm, you're not doing any good. Veterinarians can also address how long you'll have to wait before eating eggs from your treated chickens.

External Parasites

The good thing about external parasites is that you can see them and find a small sense of relief in knowing exactly what you're combating. Once you know what you've got, you can take appropriate steps to eradicate the problem. Let's start with the three types of mites that may affect your backyard flock.

Scaly Leg Mites: These microscopic mites live under the scales on your chicken's legs and feet. If left unchecked, they can cause long-term mobility issues and deformity. You can tell that your chicken has scaly leg mites when its scales are raised and swollen. Its legs and feet will be thick and scabby.

Northern Fowl Mites: When you lift your bird's feathers you can see this mite mainly around the vent, but also in thickly feathered areas. These mites feed on your chickens' skin, feathers, and blood. They spend their lives entirely on the bird and are active through warm and cold weather. They can be found on your birds in the daytime and nighttime. The life cycle from egg to maturity is less than a week, so they reproduce at rocket speed.

Red Mites: Red mites are similar to northern fowl mites since they feed on your chicken and can be found around the vent and in thickly feathered areas. They are active in warm weather, but they do not spend their entire lives on your chicken. They like to hide out in the cracks and crevices of your coop during the daytime and then come out at night to feast on your chickens. You can find these mites by closely inspecting the darker areas of your coop; you'll see clusters of these red mites hanging out together. You can also inspect your chickens at night to find these mites active. Head out to the coop with a flashlight after dark and check on your roosting birds. The life cycle of red mites is 10 days from egg to mature adult.

In addition to mites there are two other common external parasites that affect chickens.

Lice: The chicken body louse is the most common species seen in chickens. It's a flat insect that has a whitish to yellowish color. Lice lay clusters of white eggs at the base of chicken feathers. They can often be found in hard-to-groom areas like under the wings.

Tip

Do not fear if you find a louse or two has made its way to you. Chicken body lice do not feed on human blood and will jump off you as soon as they figure out you are not providing food.

Sticktight Flea: Sticktight fleas are brown and flattened like normal fleas that we encounter with other pets, but they burrow themselves into a chicken's skin and can be difficult, if not impossible, to remove. They are often noticed on a chicken's face including the comb and wattle. These fleas will also embed themselves into your other pets, such as cats and dogs, and have been known to embed in humans too.

Controlling and Clearing Infestations

External parasites reproduce and mature quickly. They can easily become a large infestation if chicken keepers are not diligent about checking their birds and watching their behavior. If you have an infestation, your birds may not want to use your coop. They won't use the nest boxes, go inside to eat and drink, or roost at night. You'll see your birds preening a lot and irritably biting at their feathers. You will see a loss of appetite and a decrease in egg production.

Parasites can be spread to your flock by wild birds and rodents, poor biosecurity where the mites come in on clothing and shoes you have worn in other coops, and by adding already-infected birds to your flock.

Chicken keepers vary on ways to control northern fowl mites, red mites, and the chicken body louse. In any case, you should thoroughly clean your coop and change the bedding during and after treatment since immature and dormant parasites can live there. Adding some cedar chips to your bedding on the floor and nest boxes can help. Diatomaceous earth (DE) can be added to the floor of the coop, nest boxes, and rubbed into the roosts. DE can also be added to your chicken's dust bath for treatment and as a preventative measure. (See dust baths in Chapter 9.) You can dust your coop with a conventional pesticide. But a word of caution: Be sure what you use is approved for poultry and labeled for the correct parasite. For instance, lice treatment may not work on mites and vice versa. Consider consulting your veterinarian, as there are medicines that can be prescribed and your veterinarian can help determine what pest you are battling.

To treat scaly leg mites, clean the coop, concentrating on the roost bars. Soak your birds' legs and feet in warm water to remove dead scales and skin. Then slather the legs and feet in petroleum jelly or rub them in linseed oil. You will need to keep coating your bird's legs and feet daily for two weeks to completely break the life cycle. In extreme cases, there are medicines that can be prescribed by a veterinarian.

For sticktight fleas, consider applying a conventional pesticide to the affected areas by using a cotton ball with the pesticide on it. That will make it easier to avoid vulnerable spots like the eyes.

Wound Care

It's important to remember that wounds can happen at any time, and a little planning ahead can make all the difference. Having emergency supplies on hand, such as latex gloves and tools to clean and protect the wound, is crucial. (See page 183 for a list of supplies.) It's also a good idea to know ahead of time where in your barn or house you can take your wounded bird for treatment and safety, as you'll need to keep it away from the rest of the flock. That way you're not in a panic trying to improvise as you go.

Caring for a wounded chicken can be scary but being prepared can ease your mind and the chicken's discomfort. From personal experience,

Little Muff as she was found after being pecked by a rooster and before her wound was treated.

there are two things that I really dislike about caring for chicken wounds: One is that you always want your chickens to be well, and it's hard to know they are not. The second is the smell of chicken blood, which is a smell that stays with you. In that regard, treating a wounded chicken is kind of like caring for a sick child: you have to get over any fears and squeamishness and do what is best.

The good news is that chickens are perfectly adapted for taking care of their own wounds. This is not to say you should leave a chicken wound alone, but rest assured a chicken's body is doing everything it can to heal itself. The average body temperature of a chicken is 103 to 105° Fahrenheit. This definitely helps fight infection.

Treating a Wound

If you have an injured chicken, start by taking your chicken to your emergency area so you can administer first aid without distractions. You may see a lot of blood depending on the type of wound. You're either going to find an open, gushing wound or a wound that's been dried and scabbed over. If the wound is gushing, then it's important to get the bleeding stopped. A chicken's body should take care of this by forming a scab, but you may need to help it a little by adding some cornstarch.

Then, assess the wound and see if there's any dirt and debris that needs to be removed. You can use a sterile saline solution to clean the wound and the surrounding area. That will give you a better idea of the situation. If you found the chicken with an existing scab, I do not recommend removing the scab unless you can see dirt and debris under it. Hot, running blood usually cleans a wound and the scab protects it. This may sound gross, but you can work gently, using the saline solution, to remove layers of scabbing without removing the base scab.

Once you've got the bleeding stopped and the wound is clean, it's a good idea to spray it with a poultry wound care solution or rinse it with hydrogen peroxide. Then it's good to add some ointment to keep the wound protected and to keep it from drying out too much as it heals. You can use an antibiotic ointment such as Neosporin or even petroleum jelly. If the wound is in a spot where it will continually get dirty, then it's a good idea to wrap it. But I find chickens don't like wrapping and will go to great lengths to remove

Little Muff healing nicely and growing new feathers.

The final result: Little Muff is a beautiful Easter Egger chicken once again.

the wrapping. So, I try to avoid it unless it's absolutely necessary. The wound should be monitored daily and cleaned and kept moist as it heals.

> ### Note
> If it's a deep wound that needs stitches, it's best to find a veterinarian who can handle the job. Also, if the wound becomes infected, a veterinarian can administer antibiotics.

Returning the bird to the flock is a judgment call. Chickens are attracted to the sight and smell of blood. They will peck a wound and injure the bird further. If the wound is small and in a hidden place, such as under a wing, then the bird can usually be returned to the flock. Also if the injury is securely wrapped and the bird is otherwise healthy, it may be able to roam with its flock mates. However, if the injury and trauma has been severe or it's difficult to hide from the flock, then it's best to keep the bird isolated until it has fully recovered.

Flystrike (Myiasis)

This is probably the most disgusting problem a chicken owner can face. Honestly, it will make you question ever having chickens at all. Flystrike is not caused by an ordinary house fly, it's caused by one of three specific flies; bot flies, blow flies, or screw flies.

These flies need a place to lay their eggs and food for the growing maggots. They can be attracted to your chickens because of a small unhealed cut or a butt that's covered in droppings. Once they find the perfect spot, they lay their eggs in the dropping covered feathers or the open cut. The eggs hatch quickly in less than 24 hours. Once the small maggots start to hatch, they tunnel through the skin causing it to die.

Since chickens are covered in lots of feathers, a chicken keeper may not notice a case of flystrike at first. There is a distinct odor when chickens are suffering flystrike and that may be the first thing you notice. You may also notice dark, blood type fluid gathering on the chicken's feathers and possibly running down the legs. Immediately inspect your chicken closely. When you find the flystrike area, you'll see an open wound with hundreds of maggots crawling around. This will be a cringe-worthy moment. And it requires immediate attention.

A chicken suffering from flystrike is in a lot of pain. If you have quick access to a veterinarian, they can treat your chicken and help ease the chicken's discomfort through pain killers and anesthesia. If you can't get the bird to a vet quickly, then you'll need to clean the wound yourself. Don sterile gloves and then carefully rinse the site to start washing away maggots. The key to success is to get rid of the maggots. This can be a hard task as the maggots really want to stay on the bird. Soaking the affected area in a bath can help you see the wound better and get rid of many of the maggots.

After you get rid of all the maggots, spray the wound with a wound care spray like Vetericyn. Keep your bird isolated from the rest of the flock. The maggots won't spread from bird to bird but the other birds may peck at the large, open wound. Don't cover the wound, let it air-dry and spray it with the Vetericyn. Check your bird every few hours to see if new batches of maggots have hatched and are doing damage again. Remove them as they appear. Keep your bird away from flies, comfortable, and supplied with plenty of food and water.

Once you get to the point where there are no more maggot hatches and they are completely gone, the focus turns to healing the infected site. Chickens can be remarkable in what they survive and their speedy recoveries. A veterinarian can prescribe antibiotics and help you choose the best course of action. But if your chicken continues to suffer from infection or wounds that are just too severe, then it may be the best choice to humanely put it out of its misery.

While you can't guarantee 100 percent prevention of flystrike, there are some ways to reduce the chances of flystrike. First, keep your coop clean and as free of flies as possible. If your chicken has a butt that's covered in droppings, bathe your chicken to remove the feces and keep the area clean. You can also trim badly soiled feathers. Treat any open wounds promptly and make sure they heal completely. Keep a close eye on your chickens, especially during the warm months when fly populations are at their highest. While you may not immediately see the wound, your chicken may not be as active or eating and drinking as much as usual. This isn't a common

Q&A: Chicken (and Egg) Health

There was a worm when I cracked open my chicken's egg. Will I get worms if I eat the egg? Is my chicken sick?

Sometimes a hen with worms can lay an egg with a worm in it. This is gross if you find it while you're cooking breakfast. While you can safely eat the egg without worrying about getting worms (chicken worms don't like the insides of humans), I would just get rid of that egg. You should watch your chicken for signs her worm load is getting too large and then treat her accordingly.

Does a chicken that's not contagious, but recovering from an injury, need to be kept with other chickens? Don't they get lonely?

During recovery, I have been known to add a docile chicken in with the recuperating chicken for companionship. I have done this successfully after the initial trauma is over and the bird is starting to heal. I think it helps the bird to integrate back into the flock with a friend. There's strength in numbers. Plus, it makes them more comfortable while they're away from the flock.

Do chickens get sick a lot?

Chickens are hardy animals, and most chicken keepers that practice good animal husbandry will not face many diseases on a regular basis. With that said, as flocks age they become more susceptible to disease and reproductive issues. So it's good to be familiar with potential problems.

My chicken died but I don't know why. How can I find out?

If you are unsure of why a chicken died and want to find out an exact cause, you can send your bird for a necropsy. Some veterinarians will do this, but more likely you'll need to find a place that can handle chicken necropsies. Your local extension office should be able to direct you to the right place.

problem, but when it happens you need to act quickly.

Treating Bumblefoot

Bumblefoot is an infection in the soft tissue pad of a chicken's foot or toe. It starts with a minor injury like a bruise or a small cut. Roost bars that are poorly designed or too high are often cited as a major cause of bumblefoot. It can also be caused as chickens are free ranging and accidentally get a cut or scrape or if chickens spend time on rough surfaces like gravel and concrete. Wet bedding and litter can increase the chances of this happening. If your chickens are getting bumblefoot often, it's a good idea to assess the height of your roost bars, the depth of your bedding, and the cleanliness of your coop.

Healthy chickens are active chickens! They are often foraging for good things to eat or finding the perfect spot for dust bathing, sun bathing, and preening.

Also check the areas where they spend the most time for sharp objects like nails that may be sticking out.

You'll likely first notice something is wrong with your chicken as it starts favoring one leg and limping. When you see these signs, take a look at the bottom of your chicken's foot. You'll see a bulge and possibly the start of an abscess if it's bumblefoot. If you catch this early, then you can clean the affected area, soak it in a warm water bath with some Epsom salts, and apply an antibacterial spray or ointment. Then wrap the area to keep it clean. Depending on the time of year and activity of your chicken, you may have to isolate the bird to keep the wound clean.

If you don't see improvement and the wound is still festering, then you'll have to remove the infection by removing the infected core

CHAPTER 5

that's just under the scab. It's best to seek a veterinarian's assistance if you're uncomfortable or squeamish. The good thing about bumblefoot is that it's a pretty common condition for birds of all kinds. So if you can find a veterinarian that treats pet birds, chances are they can help you.

If you try treating bumblefoot on your own, then first soak the chicken's foot in warm water with Epsom salts. Once the scabby area is soft, gently peel it away and push the skin aside to remove the infected core. Be warned, the core can be a lot bigger than you initially think it will be. Use tweezers to get all of the core and the pus out. If the wound is deep, you may need to use a scalpel to open more area. Rinse out the wound with saline solution or a poultry wound wash like Vetericyn. Pack the wound with Neosporin and then wrap it with gauze and tape to hold it.

Tip

Always wear gloves when dealing with bumblefoot so no infection is transferred to you.

Keep the bird in a clean environment and check the wound frequently, soaking and cleaning as necessary. Don't let the bird perch until the wound is completely healed. Seek out a veterinarian if there is no improvement.

On the surface it can seem easiest to treat bumblefoot on your own. But there are drawbacks to consider before making that choice. For one, you can cause serious damage through untrained cutting and treatment. Two, while it may seem necessary to remove the core, sometimes it's best to leave it and let it come out on its own. And, if you don't protect the foot properly, the bird can end up with a more serious infection.

CHAPTER 6
Predators

With my first flock, we hand-picked all our baby chicks, we got to know them, named them all and fell in love with chicken keeping. We built our coop by hand, making sure to make it as sturdy and predator proof as possible. We own almost 13 acres, but decided to fence in about 2 acres so our chickens could free range safely all day every day. Things were working well.

The last memory I have of that entire flock alive was visiting them in the afternoon and giving them a snack. Their afternoon snack had become a ritual; they looked forward to it and I did too! I'd call them and then sit back and watch them all waddle-run to me in the way that only chickens can do. I'd talk to them and they'd talk to me. Sometimes I'd bring my kids and the dog outside, and we'd all play for a while. Thankfully, that day they stayed inside.

About an hour after the chickens' snack was finished, my husband came home from work and asked why there was a chicken laying dead on the driveway. My heart dropped. I walked outside and immediately saw my kids' favorite Easter Egger ripped to pieces by our mud room door. I walked a little farther and saw the White Leghorn my husband had mentioned dead on the driveway. As I walked around to our backyard, I saw a dog flash by with a New Hampshire hanging out of its mouth. I started running then and what I saw in the rest of my backyard was something out of a horror film. There were dead chickens everywhere! At first, I stood there and screamed. My husband made his way back to me quickly. My kids were on our deck crying. We were all completely unprepared for the tragedy

Here you can see my first flock before the devastating feral dog attack.

that had just taken place. We had heard nothing inside the house.

Finally I snapped back to reality, stopped screaming, and started to look for survivors. I found some chickens had found a way to hide. They were scared and wary about coming out from their hiding places, but I enticed them to me and got them into the coop with treats. I also found an Easter Egger around the front of the house hiding and hurt under a bush. I know a New Hampshire was hanging around in the woods, but I never could get her to return. All in all we had nine dead chickens, one chicken that flew away, two hurt chickens that required immediate wound care, and seven unharmed but totally stressed-out chickens.

Predators are a fact of life when raising chickens. We have to remember that chickens are prey animals and make a tasty dinner for humans and wildlife alike. It may be cliché, but where there's a will, there's a way. Predators are diligent and they are resourceful. Predators can't just stop by the local grocery store and pick up a meal. They've got to work for every bit of food they get. Your chicken coop is the next best thing to a grocery store for predators. There's abundant food, from your chickens themselves to the feeder filled with pellets and crumbles, if they can just get in.

It's crucial to analyze an attack, know the predators in your area and immediately work to stop the problem. Dealing with predators is an age-old problem. It's a give and take between the natural world and our need to raise our own food. And in that struggle, it's important to have lots of tools in your arsenal to win the fight.

What's Out There?

The first step in dealing with predators is to know what you might face. Predators can be broken down into three groups: flying predators, ground-dwelling predators, and domestic predators. While there are predators that are pretty much universal throughout the United States, there are also regional predators. If you're not familiar with your wildlife, it's a good idea to talk with local chicken keepers or your extension agency to determine the biggest threats near your home. Then get to know more about those animals and their habits. Once you're armed with this information, you can work smarter, not harder, to be effective in your protection efforts.

Flying Predators

Flying predators can be daunting, especially when your birds free range. There's no practical way to put up something that fences out the sky above when your birds are roaming large amounts of land. Also, if you live near a wooded area you may have nesting hawk and owl populations with young ones to feed.

Many folks believe keeping chickens that are not white, and more camouflaged in color, will make it harder for flying predators to spot them. I'm not sure how accurate that is considering that flying predators have excellent eyesight and are used to catching prey that is camouflaged. That said, the only chicken we've ever had become a victim of a hawk attack was a White Leghorn. There's certainly no harm in following

A red-tailed hawk soars high overhead.

this practice and you'll have a beautiful flock to boot.

Depending on where you live, there will be various flying predators in residence. Then, twice a year, you will have higher numbers of flying predators as some migrate from north to south and vice versa. While you do have to be diligent on a daily basis, it's important to keep a watchful eye out during the high traffic spring and fall migrations. Also, it's good to know if you're in an area that is a popular flight path during migration. Your traffic will be even greater.

> ## Note
> Birds of prey, including hawks, eagles, and owls, are protected. This means it is illegal to harm or kill them.

As for your local resident chicken hunters, there are some common culprits. In the United

Q&A: Predators

What behaviors should you look for that chickens know a predator is around?

Predators are stealthy. After all, that's what predators are counting on for success. On the other hand, don't discount chicken senses. Chickens are prey animals, and they are always on the alert. They have good eyesight and hearing, and they're usually aware of their surroundings. If you have a rooster in your flock he will sound the alarm by calling loudly and gathering the flock for protection. In an all-hen flock, the lead hen can take on this role. Hens themselves will vocalize loudly if frightened or scared. Chickens outside the coop at dusk and during the night will either perch or lay down and remain completely still so they do not attract unwanted attention.

Are there any signs to watch for that show you predators have been trying to get into a coop?

It's a good idea to give your chicken coop a once-over each day to check for signs of predators. You may find scratch marks or bits of hardware cloth that are being loosened. In wet soil or in snow, you may see predator tracks. That will give you a good idea of the nighttime traffic around your coop and exactly who is visiting.

Can chickens experience adverse effects living in coops that have persistent predator problems?

Lots of predator activity stresses chickens. It can cause chickens to pile onto each other for protection. It can cause a decrease or complete stoppage in laying eggs. It can also lead to illness in the flock as a stressed body is more open to disease. It's important that chickens feel comfortable and safe in their surroundings.

States there are three species known by the old moniker of chicken hawk: the red-tailed hawk, sharp-shinned hawk, and Cooper's hawk. In many areas, bald eagles and golden eagles are becoming increasingly common. Owls such as the great horned owl and barred owl are not often seen, but can be heard calling to their mates during the night.

There aren't a lot of telltale signs of attack left behind by a flying predator. In fact, many times you'll just find birds missing from the flock. The most common sign would be scattered feathers and you may find scattered body parts such as a head or tail. The remains are usually seen because birds of prey will often take their kill to a nearby tree where they'll eat it.

There are differences in feeding patterns, what is eaten, and times of activity among these birds, so it's important to know the habits of the most common birds of prey.

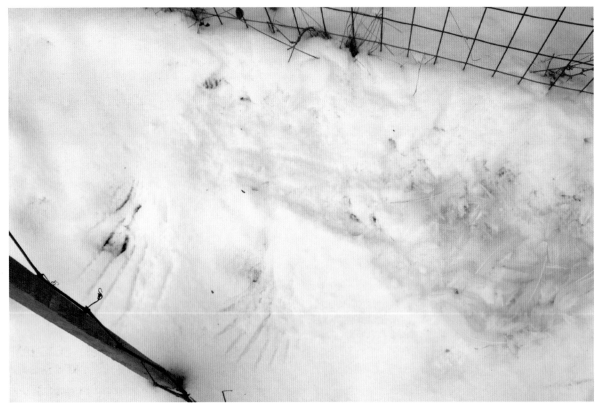

Hawk wings left imprints in the snow along with Leghorn feathers and signs of a scuffle. Lucky for the Leghorn, she came away unhurt.

HAWKS

Red-Tailed Hawk: This is probably the most common hawk in North America. It is a large (19 to 25 inches), beautiful bird with broad wings and a rounded tail. Red-tailed hawks can often be seen soaring high in wide circles. As they fly, you may be able to spot the reddish underside on the tail of an adult bird. Immature birds are streaked. Red-tailed hawks are active during the day. They raise one brood of one to five chicks in the spring. Also known as chicken hawks, they rarely hunt standard-size chickens and have a varied diet. They will take a chicken if the opportunity presents itself but the bulk of their diet consists of mammals, such as voles, mice, rabbits, and squirrels. They also hunt birds such as starlings and blackbirds. They will hunt birds up to pheasant size so smaller birds in a chicken flock make more of a target than larger birds.

Cooper's Hawk: This is a medium-sized (14 to 20 inches), long-tailed woodland hawk. A skillful flyer, Cooper's hawks can easily navigate the tree canopy to find prey and will frequently hunt along the edge of woodlands. They are active during the day. They raise one brood of two to six chicks in the spring. Cooper's hawks mainly hunt birds, especially medium-sized birds like robins, jays, flickers, pheasants, and chickens. They will also eat small mammals like chipmunks, mice, squirrels, and bats.

Sharp-Shinned Hawk: This is a small-sized (10 to 14 inches) long-tailed woodland hawk that is sometimes mistaken for a Cooper's hawk since there is some overlap in size. This is the smallest hawk in North America and is a fast flyer that's often seen on its fall migration south. They are active during the day. They raise one brood of three to eight chicks each spring. This is a pursuit hunter that will often dive and surprise its prey and can take birds in air or on the ground. Songbirds make up 90 percent of their diet. Birds that are robin-sized or smaller are at most risk, although an attack on a larger bird cannot be ruled out.

EAGLES

Bald Eagle: The national symbol of the United States, the bald eagle is perhaps the most identifiable of birds around thanks to its distinctive white head and white tail contrasting with its dark body. The bald eagle dwarfs most other birds in size (they grow to 30 to 43 inches). They are active during the day. They raise one brood of one to three chicks each spring. Bald eagles are most commonly found near water sources such as rivers, lakes, streams, marshes, and the coastline. Their preferred diet is fish, although they will take mammals and birds such as ducks when the opportunity arises.

Golden Eagle: This majestic bird is one of the largest in North America (30 to 40 inches). They are mostly found in the western half of the United States. Adults have a golden sheen on the back of the head and neck. They are active during the day. They raise one brood of one to three eggs in spring. Golden Eagles mainly feed on small-to medium-sized mammals including rabbits, marmots, and prairie dogs. They are capable of hunting larger mammals such as deer, coyotes, badgers, and bobcats and larger birds such as swans and cranes.

OWLS

Great Horned Owl: One of the most common owls in North America, the great horned owl is comfortable in many climes and habitats, including backyards. This is a large (18 to 25 inches), thick-bodied owl with two tufts on its head. They are active at night but can be seen at dawn or dusk. They raise one brood of one to three chicks in late winter or spring, nesting later in the south. Great horned owls are aggressive hunters that have even been seen hunting in the daytime. Their diet is diverse in species and size. They will eat insects as well as mammals and birds, including other birds of prey and even other owls. They will typically spot prey from a perch and then pursue it. But they can also walk on the ground to stalk prey.

Barred Owl: The call of a barred owl is a distinctive and easily identifiable chant that sounds like "who cooks for you, who cooks for you-all." This stocky (17 to 21 inches) woodland owl has beautiful white-and-brown striped plumage.

Many times all that's left behind from a predator attack is a pile of feathers.

They are active at night but can be seen at dawn or dusk. They raise one brood of one to five chicks in spring. Barred owls eat a varied diet of small animals including birds up to grouse size (16 to 19 inches). They will also hunt during the day if given the opportunity and have been known to wade into shallow water to catch fish and crayfish. They swallow small prey whole and will eat larger prey in pieces, starting with the head and working through the body.

Ground-Dwelling Predators

The predator onslaught from the ground is more robust and diverse than from the air. In my experience, this is the area where chinks in your armor are exposed and quickly become huge holes. These animals vary widely in species from reptiles such as snakes to rodents such as rats to raccoons, foxes, and coyotes. Protecting your flock from these predators is an ongoing process.

Predator populations will vary from location to location. Before you do anything, it's important to work with the local extension service to figure out what you can do to dispose of predators. Many places don't allow you to kill them; others do. No-kill traps are great, but many places have strict rules about releasing the animals you catch. You don't want to run into legal trouble when dealing with predators!

Unlike with flying predators that you will not normally encounter yourself, people can more easily come into contact with ground-dwelling predators. They can carry diseases and pests that can be transmitted to humans and pets. Keep this in mind when you're dealing with these predators and use all safety measures and precautions—especially if you have young children in your home.

In general, many ground-dwelling predators are active at night but they are also active at dawn and dusk. If your chickens free range, those are critical times to be more vigilant watching your flock. You can even avoid dawn altogether by letting your flock out closer to mid-morning.

Ground-dwelling predators don't migrate like birds of prey, so they are a year-round threat. Still, there are periods of higher activity during spring when they are raising young and in fall when they are preparing for the winter weather.

Don't forget the chicken coop itself is tempting for predators that are hungry. A feeder full of chicken food can be irresistible and can bring out animals that are not normally active in the day. I once had a perplexing problem where I'd fill up the feeders in the morning and by mid-afternoon they were bare. I knew my chickens weren't eating that much; they were out roaming. Then one afternoon I went to the coop, looked up, and saw the culprit. A young raccoon was hanging from the rafters. It didn't care about my chickens. It had found an easy meal at the chicken feeder!

The following is a list of common ground-dwelling predators in North America along with their habits and signs that you can use to identify their attack.

Raccoon: Raccoons are one of the most common and relentless predators that chicken keepers face. Don't let their cute appearance fool you; raccoons can be vicious. They are most active during breeding season from March through July. They can have more than one litter each year with two to five or more kits each time. It's important to address a raccoon problem early and often because the young learn their habits from their parents. If their parents got away with robbing the chicken coop regularly, then the young will do the same thing.

Raccoons are one of the biggest threats to a flock of chickens. They are most active at night, but it is not unusual to see them out during the daytime too.

Raccoons are strong. They are also dexterous and patient. They will spend night after night working on a particular weak point, using their nimble paws to work things open.

Raccoons are mainly nocturnal but will come out during the day if given the opportunity. Signs of a raccoon attack are missing birds, carcasses with missing head and limbs, birds pulled into fencing, half eaten carcasses, and scattered feathers.

Fox: There are two species of foxes that are widespread through North America, including the red fox and gray fox. Red foxes are larger than gray foxes. They have a red top coat, but regional color variations can sometimes cause confusion telling the two foxes apart. Gray foxes have a salt-and-pepper gray appearance with patches of red on the ears, stomach, and legs. The tails of both foxes can help with identification. The red fox has a large, bushy tail with black hairs mixed with red ending with a white tip. The gray fox has black stripe running from the base of the tail ending in a black tip.

Red foxes breed early in the season with five to six kits being born February through April. Gray foxes breed a little later with four to five kits being born from April through May. Both species have one litter per year.

Foxes are nocturnal but they are opportunistic hunters. They will stake out a chicken coop and become accustomed to your routine. They know dawn and dusk are times where chickens may be at their easiest to catch and will most often take them during those times. They can also be found hunting during the day when given an opportunity.

Topping the list of relentless predators, foxes can wreak havoc on a chicken flock. A fence is no match for foxes as they will just climb it. A solid coop with a solid roof is required to keep out a fox. Foxes will also pull chickens right out of a coop if they can find an opening. If they get in and are allowed to stay, foxes can kill an entire flock and carry off as much as they can manage. Normally though, foxes will simply take a bird or two and will leave very few clues they have visited your flock. You'll find a bird will just go missing without a trace, or with just a few scattered feathers along the route where the fox dragged the bird away.

Virginia Opossum: About the size of large housecat, the opossum is North America's only marsupial, meaning it carries its undeveloped young in a pouch where they attach to a nipple for sustenance and will take shelter until they can live on their own. Opossums are largely nocturnal animals, although they can be seen in daylight occasionally.

Opossums are omnivores, which means they eat almost anything they can find. Normally they'll target easy food, like eggs in the coop or young chicks. They don't go to the extremes that many other predators do to find food. If they do get a mature chicken, they will normally not carry the chicken away; they will eat what they want on the spot. If you find a carcass with bite marks, especially to the breast, or a carcass with the abdomen or crop eaten, those can be signs that an opossum has visited.

Coyote: Although coyotes get a bad reputation, they are not a common predator of chickens.

Coyotes are becoming more and more common though, so they may step up in the predator list in years to come. But for now, coyotes tend to be wary of humans. They are nocturnal but can be seen out at dawn and dusk as they start or finish their hunting. If they come close to a house, they'll go for the easy food and leave. They will not normally take the trouble to break into a coop. If they do come across a chicken and have the opportunity to catch it, then they will eat it. There will be no signs left save a few scattered feathers.

Striped Skunk: These curious creatures with a distinctive white stripe down their bodies are

Skunks are not often a threat to adult chickens; their focus is more on chicken eggs and young chicks. Their numbers have grown in urban settings.

rarely seen. They are more often smelled when it comes to evidence of where they've been. Skunks are highly adaptable creatures and this has allowed their numbers to grow as they move into more urban settings across the United States. Chickens are not their preferred meal. They are more apt to eat chicken eggs and young chicks. They will leave behind scattered empty eggshells if they've been feasting. And their telltale odor will remain, of course.

Least Weasel: These slender creatures can often get in areas where no one else can. In fact, the least weasel is the smallest North American carnivore. Although they have a bad reputation, weasels are actually a farmer's helper as they prefer to eat farm pests like mice and rats. Weasels go about their business and are rarely ever seen since they prefer to stay away from humans and a chicken coop. But if hunger or the opportunity arises, they will make a run at your chickens. Their small size allows them to get in places that other animals can't, so it's important to use the recommended ½-inch galvanized metal hardware cloth to cover any openings. A chicken coop that's raised off the ground helps deter weasels. If you have a ground level coop, make sure there are no mice or rat tunnels into your coop since weasels can use those tunnels to travel. If a weasel attacks your chickens, it will normally leave plenty of evidence since it's likely it won't be able to fit the dead chickens out the hole it came in through. It's most likely to go for the eggs first. But weasels are attracted to movement and flapping scared chickens make an easy target. If it does kill any chickens, weasels have an interesting habit of piling the bodies in hopes of storing them for a later meal. You will also find the back of the head and neck eaten along with small bites on the bodies. Since the least weasel is a member of the same family as skunks, there will be a musky odor left behind.

Domestic Predators

While most people worry about the local wildlife, domestic predators may be the biggest threat you face. Roaming populations of dogs are a huge killer of chickens every year (as I found out the hard way). I'm not alone: Many long-time chicken keepers say dogs are the number-one predator of their chickens. This is just as true in rural areas as it is in neighborhoods. It's just as true for feral dogs that have no home as it is for pet dogs that roam among the houses in your neighborhoods. There is no way to know when dogs will show up and they can be dangerous if you're trying to protect your birds.

Unlike wild predators that are hunting for food, dogs normally kill for the fun of it, so they leave lots of clues that let chicken owners know they are responsible for the attack. You will see dismembered birds lying around but not eaten, scattered feathers, and fences that have been ripped open. Don't make the mistake of thinking that dog attacks only happen during the day. Although your pet dog may happily sleep the night away, packs of dogs can be found roaming at night.

Dogs also like eggs, so while your tame pet may not be dangerous to the flock, dogs can and will devour eggs if given the chance. If you find lots of empty broken eggshells lying scattered around the coop, that can very well be the work of dogs.

Cats are not as much of a threat to chickens as dogs. They do not typically kill fully mature large chickens. Cats are smart and they quickly learn that full-sized chickens are too big to catch. Baby chickens, on the other hand, are bite-sized. If you let chicks out to roam, make sure they are in a safe enclosure and they are supervised.

Cats, if they attack, will do so at any time of day, although you may find them more active at dawn and dusk during the warm times of year. If a cat does attack, it will often eat a chick, possibly leaving behind only feathers. Cat attacks that have not been so successful can result in bite and scratch wounds and areas of the body with missing feathers.

Keeping Your Flock Safe

Nothing is 100 percent effective in keeping your flock safe. The reality is that most chicken keepers will face some type of predator problem and experience losses. Each story is different, and many times people feel guilty. I know. I've had the same feelings. But it's important to move on and protect the chickens that are left. Many people give up raising chickens after an attack, but please don't let predators stop you. Learn from any mistakes and use the tools at your disposal. The good news is that there are many to choose from, and combined, they can be very effective.

Fake owls and hawks may seem silly, but many predators don't want to take a chance
if it appears another predator is on the scene.

Use Popular Deterrents

Get a Watchdog: Not only are the best watchdogs adept at keeping a watchful eye out, just having a dog in the yard can be a deterrent. If your dog doesn't stay out all day with the flock, then it's a good idea to vary her routine by letting her out at different times each day. Predators are smart and they catch on to routines quickly. A varied dog routine makes for a more nervous predator.

If you don't have a dog and are considering adding one to your home, there are livestock guardian dogs that are trained to protect their charges. Other purebred and mixed-breed dogs can also serve this purpose. It's a good idea to do some research on dog breeds before making a purchase. Some breeds are better suited to being around chickens and pets than others. Be sure your dog is not a threat to your chickens before assuming the responsibility of flock guardian. If you're not sure, it's a good idea to contact a trainer that can help.

Predator vs. Predator: Hawks don't like owls and vice versa. Plus they don't want two of their own kind in the same backyard. So head to your local farm supply store and pick up a fake owl or hawk. Mount it in your chicken yard and be sure to move it around to get the full effect.

Plant or Build for Cover: When chickens spot a predator, they need a place to hide. Many chicken coops are raised, so underneath the coop becomes a great hiding spot. Decks and house overhangs provide protection from weather and predators as well. Shrubs and bushes also make great hiding places. (See opposite page for more on securing your coop.)

Make a Scarecrow and Hang Shiny Objects: In the fight against aerial predators, it's good to remember they're looking down on your chicken yard from above. They know what belongs there and what doesn't. They need to think that

A scarecrow can also be a great deterrent to predators.

something is continually down below that they don't want to encounter. So press your Halloween scarecrows into service year round. Mount them around the chicken yard. They've got kind of a rustic charm, too! Oh, and grab some shiny objects like aluminum pie pans and old CDs while you're at it. Hang them from random tree branches so they move with the breeze and glint in the sun.

Enlist Male Help

Roosters Make Great Hen Protectors: Hens are pretty good at protecting themselves. A flock without a rooster usually has a lead hen or two that keeps watch. However, adding a rooster definitely steps up the level of protection. A good rooster can often be found scanning the skies for flying predators. If he sees something, he's quick to let out an alarm call and gather the hens in a protected spot. Usually the rooster will pace back and forth in front of his gathered hens keeping them safe until danger has passed. Roosters will routinely roam the edges of their territory and even follow hens that break free of their confines and wander off.

Consider Urine: Yes, this is gross. Yet scent markings are an important communication tool in the animal world. So communicate like predators do naturally. This is how wild predators set up their own invisible fence and know what's going on in their territory. If you have a male human member of your household, encourage him to go outside and urinate near the coop on a regular basis, just remember to be mindful and discreet if you have close neighbors. (Female urine doesn't work as well.) Don't do this in high-traffic areas because that's just not sanitary. Have him mark a few areas around the perimeter just like a predator would.

You can actually buy predator urine as well and will regularly find it for sale in chicken catalogs and websites. If you don't have a male in your household, this may be an option. Just make sure no animals are harmed and the urine is collected humanely.

Consider Routine

Adjust Coop Opening and Closing Times: Know the time schedule predators keep and don't put your chickens in harm's way. Avoid opening the coop ultra early and letting the birds roam right away. If you've got a covered run, let your birds start their day in the run then open that to the outside once the sun is up.

Use Alternative Strategies: Since predators do become accustomed to your prevention strategies, it's a good idea to use everything in your arsenal and mix things up. Alternative strategies can help you with a stale routine. Plug in a radio and play some music near the coop during the night. There are also predator lights that can be hung around the coop. At night they flash a red LED light that threatens predators.

Rules Are Meant to Be Broken: While we can look at general activity patterns for predators, those are just guidelines. Never underestimate a hungry animal that needs to find its next meal to survive. They'll come out at any time of day; especially if they have young to feed or are young

and newly on their own. In my neck of the woods, raccoons will sometimes find the feeder in the coop irresistible.

Use Technology: Besides sleeping all night in the coop, it's hard to know what's going on there while you're getting some rest. If your coop isn't too far from the house, you can set up a baby monitor and leave it on during the night and when you're at home. You'll hear the commotion if anything untoward is happening in the coop. You can also set up a field camera outside the coop that takes pictures based on movement. Download the pictures and you'll have your predator on film. Remote computer cams allow you to watch your coop from any location.

Secure the Coop

I once heard someone say that chicken coops shouldn't be designed to keep chickens in; rather, they should be designed to keep predators out. I think that's good advice. While a coop can be a secure place for your birds, it can also be a death trap. If birds are free ranging and they're attacked they do have some means of saving themselves. They can run, they can fly, and they can hide. But in a coop or a run with a predator, there's nowhere to go, especially at night when it's hard to see. Thankfully, there are some coop construction methods that work well and add extra protection.

First and foremost, make sure that your walls, ceiling, floor, and doors have no gaps or cracks that can allow predators access. Enclose all openings such as windows and ventilation spots with welded wire. Also, set your coop off the ground

A good rooster is always scanning the yard for predators.

to prevent predators from digging under and into the coop. If you install an enclosed and/or covered run or a coop directly on the ground, make sure to bury welded wire 6 inches down and 12 inches out parallel to the coop making an L shape with the wire. This will prevent digging predators.

While it might seem okay, do not use chicken wire for a chicken coop or run. Chicken wire has large hexagonal openings. It's flimsy and easily pulled apart. (You'd be surprised at the small openings that predators can fit through to reach your flock.) There are some instances around the coop where chicken wire can come in handy, such as building a partition or plugging up a hole by crisscrossing it over itself to form a barrier. If you

use chicken wire in those instances, be sure it's in good condition. Chicken wire is easily broken, leaving sharp pieces that can harm both you and your chickens. Never use it as flooring. In general, it's best to keep chicken wire for crafting and decorations only. Instead, use narrow-gauge welded wire. (This is also called hardware cloth.) It comes in a variety of gauges, so be sure to pick galvanized wire with no more than a ½-inch opening.

Last, don't use hooks and loops for latches. Predators, especially raccoons, can use their paws to easily work a hook and loop latch open. They can also open deadbolts. Try spring-loaded eye-hooks or a lock and key set—just make sure it's a sturdy set.

Spring-loaded latches will help safeguard your coop against prying animal paws.

Galvanized hardware cloth with a ½-inch opening can help deter predators.

CHAPTER 7

Feeding Your
Chickens

People often get chickens because of the eggs they provide. Yet one of the biggest reasons my family bought chickens is because of what they eat. You may think this sounds like a weird statement, but it's true. We needed free-range chickens and we needed them fast.

It started when we moved to the country and built our house on our farm. We were thinking of adding some livestock. Our girls were very young at the time. I took them out to the backyard sandbox to play. After uncovering it and sitting them inside, I glanced down and saw a completely black spider with no markings on her top. I'm a naturalist so I knew exactly what I was seeing. I told my oldest to step out and quickly grabbed my youngest. I caught the spider in a jar and at that point I saw the telltale red hourglass on her abdomen. My kids had been in a sandbox with a black widow spider! I was unnerved by this, but the reality is that black widows are in my area. They're pretty reclusive, so I thought this would be the end.

But then we started finding black widows everywhere, tons of them. We had an infestation! We were able to track down that the spiders had come in when we had a boulder retaining wall built. They were probably located on the black tarp rolls they used to hold the soil behind the boulders. We sprayed that wall with poison and hundreds of black widows actually came out and died on our driveway. Then we called a pest control company. But we still kept finding the spiders. My husband got bitten twice. I hated using the chemicals. And the situation was becoming dangerous. We needed another solution to this problem or we would have to move.

We decided to cancel our pest contract and chickens became our new plan. Once our flock was old enough, we let the birds free range everywhere around the outside of our house. We even opened up our garage, pulling everything out and letting the chickens roam there. It worked! From then on we saw no more black widows, and we definitely noticed a great reduction in the bug population in our backyard.

Not only do we continue to benefit from what our chickens eat, our chickens do too. They graze on grass and weeds. (Ours love white clover!) They eat insects and other animals like frogs, snakes, and even mice. Chickens, after all, are omnivores. You can tell free ranging agrees with our chickens by the eggs they lay, with deep yellow yolks and a rich creamy taste, and by the long lives they live. For us, free ranging is a chicken-keeping choice we embrace. In addition to feeding our chickens healthy snacks from the kitchen, free ranging is a great way to supplement a chicken's diet and it provides a nice savings in feed costs. During spring, summer, and fall, we notice a significant reduction in the amount of commercial feed purchased. It is offered free choice all day, but the chickens fill up as they free range and have less need for it. So, if you have a chance to let your flock eat naturally, I'd recommend it for both them and you.

There are many feeding options available for the backyard chicken keeper. I'm obviously a fan of free ranging because it's a good complement to a well-rounded diet. In this chapter we'll explore what makes a well-rounded diet, from which commercial feeds are available and when they should be fed, which treats can and cannot be given, and what supplements can help your chickens stay healthy.

How Chickens Eat and Digest

Watching chickens eat and drink can be a relaxing and enjoyable experience. They are so earnest while eating, especially if they've been given a treat. And when they drink, it can be comical to watch chickens dip their beaks into the water and then tip their heads back to swallow. But where does the food and water go and how is it digested?

Let's start with drinking. A chicken's tongue is not completely effective for pushing water back. When we swallow, we close our mouths and let our throats do the work. A chicken gathers water in its mouth and tilts its head up to let the water go down its throat. Chickens have a hole in the roof of their mouth called the choana that connects to the nasal passages. As the chicken swallows, the choana closes so that water doesn't come out the nose.

As for food, you've heard the saying, "rare as hen's teeth." That's because chickens have no teeth so they can't chew their food. They eat their food whole, breaking it up as best they can with their beaks before swallowing. Chickens do have salivary glands. The saliva helps to wet

It's always interesting to watch a chicken drink. They take water into their beak and then tip their head back to swallow properly.

the food, making it easier to swallow (and it gets the digestive process started through an enzyme it contains called amylase which begins to break down the food). From there, the food moves through the esophagus and into the crop, which is like a holding tank for food that the chicken has eaten.

Food then moves through the proventriculus, where it is further broken down through chemical secretions and then into the gizzard (ventriculus). This organ serves as a hen's teeth. For the gizzard to work properly, it needs grit to grind the food into a soupy mixture that's easy to digest. The grit and the muscular contractions of the gizzard replace the act of chewing. If your hens never eat anything other than commercial feed, it's likely they don't need grit. The reality is that your chickens are going to eat more than commercial feed and will need grit throughout their lives. Free range chickens usually get grit as they forage. Confined chickens are less likely to find grit, so you can buy grit and offer it free choice. If you're not sure if your chickens are getting enough grit, go ahead and offer it free choice. That way you're covered. Grit comes in different sizes according to the size of your chickens: smaller for baby chicks and bigger for mature birds.

From the gizzard, food moves through the rest of the digestive system including the small intestine where nutrients are further broken down through liver bile and pancreatic enzymes and then absorbed. Ceca break down the remaining food even more before it moves into the large intestine. The last stop before excretion is the cloacae. This is an intersection where the digestive, urinary, and reproductive systems meet. Since chickens have no bladder and don't urinate, the uric acid from the kidneys mixes with the chicken's feces here to form the white cap that's often seen on a chicken's droppings. Chickens excrete their waste through the vent.

Feeds

Today's commercial chicken feeds are well-balanced for a backyard flock and should make up the bulk of your chicken's diet. There are so many choices at feed stores, from big-name brands to regional brands and locally milled feeds. As your flock expands, it's good to know what makes a healthy feed and how to use it for all your flock needs.

Chicken History

I found this beautiful Kraft chicken sign at a yard sale. The vibrant colors caught my eye first, the chicken shape caught my eye second, and the logo from Kraft proclaiming, "I Get the Milk Bank Boost from Pex," sealed the deal. The decorative value is personal, but the historical perspective on chicken feeds is fascinating.

It turns out my sign is one of three designs for metal signs made for the Kraft Foods Agricultural Division in Chicago, in the late 1950s to 1960s. They were produced by the Stout Sign Company out of St. Louis, Missouri. There is a pig-shaped sign that says, "I Get the Milk Bank Boost from Kraylets" and also a calf shaped sign that says, "I Get the Milk Bank Boost from Kaff-A"—all are embossed metal and have the same yellow, green, and red colors.

Farm signs were a staple for salesmen back in the day. Without computers and technology, these signs were an early form of calling cards. I was lucky enough to find a copy of *The National Future Farmer* magazine from the Future Farmers of America dated October/November 1963, where they had an advertisement that explained Milk-Bank Booster feeds.

It turns out that these feed boosters were made with milk by-products and were meant to help animals get more nutritional power and balance from their rations. According to the ad, these feeds produced faster, more economical gains, better health and resistance to stress, and better productivity. All this was accomplished by adding the extra nutrition of milk by-products to the ration and by unlocking more nutrition from the other elements of the ration.

"Milk-Bank Feed Boosters are storehouses or banks for the key nutrients of milk: lactalbumin protein, milk sugar, vitamins, minerals and important growth factors—elements not found in ordinary grain rations, pasture, or roughage." Kraft Pex came in bagged form for everyday feed and in block form like today's poultry blocks.

Starter/Grower Feed

Starter feed should be used for egg-laying flocks until your chicks reach 16 to 18 weeks of age. If you are raising chickens for meat or have a mixed flock with chickens and other poultry, like turkeys and waterfowl, then you should start your birds on a flock-raiser feed that has protein for added energy and efficient weight gain.

When you begin keeping chickens, chick starter is a natural. But once your flock expands, many people get confused about what to feed their flock since they've got layers and younger chickens combined—especially if you've used a broody to hatch chicks and they're already a part of the flock.

What's important to know is that layer feed is dangerous to chicks because the high levels of calcium it contains can cause kidney damage. Chick starter won't hurt anyone in the flock so it should be given as the feed of choice until the chicks are old enough to eat layer feed.

It's interesting to compare the thoughts of chicken owners from 1963 to over fifty years later. Today, many chicken owners wonder whether it's OK to give their chickens milk. Yet if you're lucky enough to keep chickens and live on a dairy farm or have a family cow, you've likely fed them some of the extras and know that chickens love milk. Milk's nutrients haven't changed over the years; it still has proteins, carbohydrates, vitamins, minerals, and fats that are beneficial to chicken health. Chickens are not lactose intolerant. In fact, their bodies produce small amounts of lactase, which is necessary to breaking down the lactose in milk, but too much milk can cause diarrhea. Feeding small amounts of milk or its forms such as cottage cheese, yogurt, buttermilk, and cottage cheese can be beneficial. Whey, the liquid that's expelled during the cheese-making process, is also a great option if you have access to it.

Kraft Pex proudly advertises chicken feed containing milk byproducts to add protein and balance to daily feed rations.

Calcium can be provided free choice for the laying hens.

The big choice with starter feeds is whether you feed one that's medicated or not. This is a hotly debated subject in the chicken world and it centers on what is widely regarded as the number-one killer of baby chicks, coccidiosis. (This highly-contagious parasitic disease kills quickly and moves through a flock at high speed.)

It's important to understand the difference between the feeds and make a choice that's comfortable for you. Plain starter feeds contain no medicines, just feed. If your chicks have been vaccinated for coccidiosis, then this is the feed for you. Medicated starter feeds usually contain amprolium, which is a coccidiostat. It reduces the growth of coccidian oocysts and lets the unvaccinated chicks get past a vulnerable time as they grow into adults and develop their immunities. Some folks are strongly against giving any type of medicine to their chicks. They prefer a natural approach and say that if you keep the brooder clean, there's no need to worry. Others say there's no need to use preventative measures and choose to treat for the problem if it arises. If you choose to feed a medicated starter, then it's best to keep feeding it until your chicks graduate to layer feed.

Layer Feed

After 18 weeks, your chicks are grown and ready to move into their egg-laying cycle. They need a little less protein and more calcium to support healthy egg development. They are ready to start laying eggs and layer feed should be the food of choice. Layer feed has slightly less protein and includes calcium for strong eggshells.

If you're expanding your flock, then you likely already use a layer feed. These feeds are continually being tweaked by the major feed companies to provide maximum nutrition for the exact needs of your flock. Nowadays, you can find feed additives such as marigold extract for a richer yellow egg yolk, prebiotics and probiotics, omega-3 fatty acids, and extra calcium for strong eggshells. No matter what brand you choose, there are two main feed forms: pellets and crumbles. They both have the same nutritional value, but pellets may reduce waste around the feeder as food gets dropped. Crumbles are said to be messier. In the opinion of my flock, they prefer crumbles. There is a third, less popular and easy-to-find form of feed called

These chicks are gathered around a feeder that's been raised on a landscaping block to prevent spillage.

mash. This usually comes directly from your local feed mills and is a less formed, more powdery crumble. If you can find a good local mill, chicken mash can be an ultra-fresh food for your flock.

Feed Supplements

Commercial feeds, nutritious treats, and free ranging generally provide a chicken with a proper diet, but there are some supplements that you may want to provide to help with healthy egg laying, cleanliness, and overall good health.

Calcium: Laying hens use the calcium from their bodies to form eggs. Almost half the calcium a hen needs to form an eggshell comes from her bones. Hens that can't get enough calcium in their diet can easily deplete their bone supply and have weak bones.

Fun Fact

The calcium a hen gets for her eggs comes from her medullary bones, which include the tibia, femur, pubic bone, ribs, ulna, toe bones, and scapula. These bones are found in both hens and roosters and serve the same structural purpose in both sexes. It's only in the hens where these bones serve a dual purpose supplying calcium.

Calcium in commercial layer feeds usually comes from two different sources, limestone and oyster shell. Both break down and are absorbed through the digestion process. It's also important to offer calcium free choice to your hens. They will eat what they need. You can buy oyster shells, but many chickens, including my own, just don't like them. A great way to offer calcium is to feed them back their own shells. Save the shells after using the eggs. Rinse them and then microwave them for a few seconds to make them crunchy. Crumble them up and offer them in a separate container so the chickens can eat them as they need.

If you have a rooster or non-laying hens in your flock, they do not need extra calcium for egg production. It's hard to feed a mixed flock with birds of all shapes and sizes different foods, however, and maintain that separation. Most backyard chicken owners feed a layer feed to the whole flock and find no adverse results. But if you have the ability to feed separately you can feed a general grower feed to your roosters and non-laying hens. This works well if you have a rooster bachelor pad or an older flock with only non-laying hens. You can also feed a general grower feed to a flock with roosters and laying hens, just be sure to offer plenty of calcium free choice to maintain good egg production.

Apple Cider Vinegar: Apple cider vinegar (ACV) is a must-have for a backyard flock and can easily be found at the grocery store. You can use either pasteurized or raw, organic ACV with the "mother" in the bottle. If you use the raw ACV, be sure to store it in a cool, dry place with the lid properly tightened, to prevent bacterial contamination.

Perhaps ACV's most important benefit is that it helps a flock with cleanliness. If you add ACV to your chickens' water, it will keep your container clear of the slime that can build up when water sits stagnant. Slime is a Petri dish for disease. You can wipe it each time you fill the water, but it can be difficult to reach every place the water touches, and without ACV the slime will keep returning. ACV works by making the water slightly acidic, which is inhospitable to the slime. ACV also helps to increase a chicken's immune system and increase your bird's mineral and calcium absorption. It's helpful to promoting a good mucus flow.

Dosage: Add 1 tablespoon per gallon of water. (Double this amount if you have hard water.) You can use it in two different time frames: either once or twice a week or for one week straight per month. For chicks, a splash in their water keeps the container clean and gets them used to the taste.

> ### Tip
> Do not use ACV with metal or galvanized containers. It will corrode the metal. It's best to use it with plastic or ceramic containers.

Garlic: Garlic is good for humans and it's good for chickens too. You can purchase organic garlic at the grocery store year-round or plant it in the fall for a late spring/early summer harvest. Garlic is a bulb, so you plant it as you would decorative bulbs in your garden, just not as deep (only about 2 inches). When the time comes, you'll know your garlic is ready to harvest because the shoots will turn yellowish/brown and droop. You can also plant garlic cloves in a container in spring and then harvest the shoots as they grow and eat them as you would chives.

Feeding clean, crushed eggshells is an easy way to give your chickens extra calcium.

Garlic used in chicken keeping provides a natural antibiotic and immune system booster, and aids in respiratory health. It can be a laying stimulant. Garlic fights external parasites like lice, mites, and ticks since chickens that have eaten garlic don't taste very good. Last but not least, garlic is a natural way to fight worms and keep your chickens clear of worm infestations.

Garlic can be given to chickens in many ways. In powdered form, it can be added to their feed. Just sprinkle it on their feed, mix it a little and you're done. You can also chop it up or crush the cloves and offer it free choice in a separate dish, or float it in their water. This works well, but be careful. After a day, the water can develop a strong taste that chickens dislike. This can lead

Q&A: Feeding Your Chickens

What treats can chicks in a brooder eat?

Chicks can eat treats. But remember that if you feed them treats, they need chick-sized grit or dirt so they can properly digest their fun food. Some good treats are scrambled eggs, cut-up grapes, and oatmeal. Do not feed chicks oyster shells.

Will feeding garlic make my eggs taste bad?

Eggs from chickens that eat garlic don't have a "garlicky" taste. Some people actually say they prefer the taste of eggs from chickens that are fed garlic; they say the eggs have a milder taste.

How should I store my commercial feed?

Feed should be stored in tightly lidded containers to keep out pests. It should be kept in a cool, dry location to prevent spoilage.

Is it a good idea to make my own feed? Can I switch feeds?

It can be tempting to try and make your own feed, but it's an uphill task to get all the formulations right so your birds are receiving the best nutrition possible. It's best to use a reputable feed as the bulk of your flock's diet and then supplement with nutritious treats (not to exceed 10 percent of the chickens' diet). When you're switching feeds, do it gradually so your birds have a chance to adjust. That's why it's a good idea not to let your stored feed get down to the last crumb and then head to the feed store. You could find your store is out of the food you normally use.

Can I feed my chickens scratch grains instead of regular feed?

Scratch grains are considered a treat. They do not have the balanced nutrition of a commercial feed. They should be given at the same 10 percent ratio as for treats.

---

to them drinking less water. So, if you add garlic to water, change it daily and consider offering a separate feeder with plain water too.

From the Kitchen to Your Flock

Treats from the kitchen are a great way to recycle your leftovers. They are fun for both you and your flock. Chickens enjoy variety and their diets can gain depth through nutritious treats. Pay attention as you're feeding chickens scraps from the kitchen; soon you'll find they have favorites and you can be sure to provide them more often. If the chickens don't enjoy a treat, you may have leftovers that should be added to the compost pile or the trash so they don't rot and make a mess. Always be on the lookout for treat opportunities! Next time you clean out your refrigerator, scrape

Common Foods That Can Be Eaten and Enjoyed by Your Flock

Apples (The seeds contain cyanide, but not in sufficient amounts to kill.)

Apricots

Asparagus (Limit the amount; it can taint the taste of eggs.)

Bananas (Do not feed the peels.)

Beets (plus greens)

Blackberries

Blueberries

Bread (Try to offer healthy bread to give your chickens the biggest bang for their buck.)

Broccoli

Brussels Sprouts

Cabbage

Cantaloupe

Carrots (plus greens)

Cereal (Try to avoid sugary cereals though.)

Cherries

Collard Greens

Corn (Most chickens especially love corn on the cob.)

Cranberries

Cucumbers

Eggs (Hard-boiled eggs are yummy. Warm scrambled eggs are perfect on a cold morning.)

Fish

Garlic

Grains and Seeds (Sunflower seeds can be given with or without the shell. You can purchase black oil sunflower seeds in bulk in the wild birds store section.)

Grapes (You can even cut grapes in half for chicks.)

Grits

Honeydew Melons

Kale

Lettuce

Meat (You can also give your flock the bones and they will pick them clean.)

Nuts (Avoid salted, seasoned, and sugared nuts.)

Oats (Cooked or raw oats are fair game.)

Parsnips

Pasta

Peaches

Pears

Peas

Plums

Pomegranate

Popcorn (Leftover movie theatre popcorn is great. Refill your bag as you leave the movie and make sure not to add butter or salt.)

Potatoes (cooked)

Pumpkins

Radishes (plus greens)

Raisins

Rice

Seafood

Seeds

Spinach (Feed sparingly as too much can interfere with calcium absorption.)

Sprouted seeds

Squash

Sweet potatoes

Tomatoes (Do not feed green tomatoes, leaves, or vines.)

Turnips

Watermelon

Yogurt (Watch the flavors though. Try to stick with plain or natural flavors like strawberry, banana, or blueberry.)

Zucchini

the dinner plates, or bring home leftovers from dinner out, why not set some aside for your birds? They'll love you for it!

Just make sure treats don't become the bulk of your chicken's diet; no more than 10 percent of their food is a good rule. This way the full nutritional impact of their commercial feed is not diluted too much.

What to Avoid

There are some foods that should *not* be fed to chickens. With that said, chickens are smart. Chickens that free range may come in contact with poisonous plants, but they learn quickly what not to eat. Many chicken keepers find that even if the foods below are offered to their flock, they will not be consumed. The exception to this is a flock that doesn't get out of its coop. They often don't learn how to distinguish what's good or not good for them and are more likely to eat toxic substances. Also it may be hard for chickens to distinguish bad items when mixed in with a large variety of treats.

AVOID FEEDING THESE ITEMS COMPLETELY

- Alcohol
- Avocado pits and skins
- Chocolate
- Citrus
- Fried foods
- Onions (Some may be included in leftover dishes. It's OK to feed these dishes as long as the onions are minimal.)
- Raw green potato peels and sprouting eyes (These contain a toxic substance called solanine.)
- Rhubarb
- Salty foods
- Sugary foods
- Undercooked or dried beans

While you're cleaning the fridge or feeding leftovers, be sure not to feed your chickens food you wouldn't eat. A general rule of thumb for deciding which treats to give your flock is if it's good for you, it's good for them. If food is moldy or rotting, then it's not good for you and not good for the chickens either. It should be transferred to the trash bin.

While a chicken can eat many of the same foods you enjoy, there are some you'll want to avoid.

CHAPTER 8

Chicken Keeping through the Seasons

CHAPTER 8

Gardening, especially herb gardening, was my first love. I can remember gardening with my mom as a kid and eagerly eating the green tomatoes as I picked them. (A feat that earned some admonishment as my mom pleaded with me to wait until they ripened.) Sitting among our sunflowers and smelling the fresh herbs my mom grew is still a cherished memory.

Once I left the house and got my own apartment, I planted wherever I could find space; a zinnia here, a pot of herbs there. My first house didn't come with an extensive herb garden, but I left it with one and took tons of cuttings along with my favorite viburnum bush.

At that point we moved to the country. We had plenty of space to plant—and to raise chickens, of course! Many people ask how I'm able to garden with chickens. My answer is that it's taken many years to figure that out. Before chickens, I had beautiful beds with vegetables and decorative plants growing together. I had an extensive herb garden that I loved to have my kids visit and taste and smell the different plants. All of my beds were regularly mulched with dark hardwood twice a year.

Then came chickens. As soon as they were old enough, I let my first flock out to roam. And anyone who knows chickens knows what I'm going to say here. My chickens completely disrespected the garden boundaries. They scratched and pecked everywhere. Mulch no longer graced my gardens and the only thing left standing were the big bushes.

I read all about how chickens are helpful to the lawn, but the hillside to my coop no longer held grass and instead was bare with dust bath holes and became a muddy, slippery mess in the rain. I tried fencing to keep the chickens out. That looked ridiculous with tons of separate fences all over the backyard. Plus, as a novice chicken owner, I had no idea chickens flew so well; especially my White Leghorns. I tried keeping my chickens in their coop and run all day, but that was just sad. So instead of fighting them, I joined them and today have a very nice yard indeed.

First, my chickens became my garden helpers. My Barred Plymouth Rocks were my favorites for garden duty. Any time I'd dig a hole or move a log or stone they were right there looking for insects and worms. They'd run over and tilt their heads sideways as they surveyed the scene for yummy treats. Once they found something, they'd start scratching and pecking with such determination. They'd practically dig my holes for me! I moved my tender plants to my front garden, which has now become a dedicated herb garden. In my yard, where the chickens roam, I planted only things that are hardy and grow fast. For instance, my pool is lined with knock out roses and butterfly bushes. Viburnum, arborvitae and barberry are impervious to chickens plus they provide great shade and protective cover from flying predators. I ditched the hardwood mulch and switched to river rocks of varying sizes. I also moved my coop down the hill to more level ground and created rocked paths to the coop along with a rocked front entrance so mud and dirt aren't tracked inside. I did fence off the hillside while I seeded new grass, but opened it after the grass fully took. I still grow herbs in the chicken areas, but not anything that's a special plant, like my lavender thyme. There I have hardy herbs like mint and

These young chicks love to flock around a lemon balm plant and nibble its leaves.

its many cousins. My vegetable gardens are raised and have fencing around them. I only open them up when it's time for the chickens to perform clean-up duty. I'm actually considering moving them entirely to a different place.

So, is gardening with chickens possible? I say yes. It can be a rewarding endeavor, but you've got to get into the mindset of the chickens. Think before you plant and be prepared to take a few losses along the way. With trial and error, your gardens and chickens can be happy and healthy and provide year-round enjoyment.

Spring

Spring is a temperamental friend to the chicken keeper. This is the time of year when you can open your coop doors wide, let some fresh air inside, and get your flock outside for some serious free ranging. It's also the time of year hens will go broody, with new chicks soon to follow. But depending on where you live, beware of extreme weather changes during this time and make your plans accordingly!

For the gardening chicken keeper, spring is a great time to assess which plants in your herb garden made it through the winter, what you'll have to replace, what annuals to buy, and what new herbs to try this season.

If you've never used herbs in your chicken keeping, they are an easy and enjoyable way to help keep your birds healthy and your coop free of pests. Herbs are being used more and more in everyday medicine and in livestock medicine as we struggle with antibiotic resistance and look to find natural ways to support our immune systems and overall health. While nothing can replace a veterinarian or well-researched and tested modern medicines, our flocks can benefit from adding herbs into their routines. Most of us already keep a few herbs around our homes for culinary enjoyment. Those same herbs can be put to work in chicken keeping. They're a win-win for both humans and chickens!

Adding herbs to your coop is an easy way to expose your chickens to helpful plants without forcing them on your birds. By tying herbs in bundles and hanging them around the coop or sprinkling herbs in nest boxes, your chickens can choose what they'd like to eat, how much they eat, and when they eat. Chickens are intuitive and they often know what can help them and what cannot. There's no exact science to this. You can start by trying out the herbs you have the most of in your garden. Or you can try to tailor the herbs to your chickens, using relaxing herbs if you know bad weather is coming or if you've got a broody in the house.

> **Tip**
>
> If you add herbs to nest boxes, make sure they are dried and not fresh. Fresh herbs can mold and decay without proper airflow.

Herbs can also be grown in the areas your chickens free range. This is another great way to give chickens the choice to add herbs to their diet. If you're starting from seed or transplanting seedlings, it's a good idea to give the plants some protective covering. There are many options, including making a small fence or cloche from chicken wire. Once your herbs are well established, you can remove the barrier. Most herbs are hardy and can usually withstand attention from your chickens. In fact, you may even find your chickens laying eggs in your herbs and routinely resting on favored plants.

Directly adding herbs to your chicken's food and water is another option, especially for flocks that don't free range. During the growing season, you can add fresh herbs directly to the feed or offer them free choice in a separate bowl. If you preserve your harvest by drying your herbs, you can add the dried herbs year round. Let your chickens decide how much or how little to eat and then discard the leftovers. For a treat that works for chickens and humans alike, you can make a spa water by steeping whole herb leaves, strain when the spa water is ready, and then serve it in an extra waterer. By offering fresh, plain water along with the herb water, your chickens can choose.

When adding herbs to your chicken keeping, remember not to make them mandatory for your hens. Start with small amounts and see if your chickens like that particular herb. If they don't, that's okay, too! Having a non-herbal option available will keep everyone happy. But with so many beneficial herbs to choose from, you're sure to find a few your chickens will like.

Note

In summer, I often mount a hanging basket filled with lemon-scented geranium from the front of my coop. In nurseries and at hardware stores, this plant is called a mosquito plant or citronella plant. It's an herb that gently repels insects if you rub the leaves to release its scent. Mosquito plants like a partially shaded spot away from the hot afternoon sun and benefit from being trimmed regularly. This is a tender perennial so it must be overwintered in cold climates or treated as an annual.

Top Six Herbs for Chickens (and Humans)

Here's a list of my six top herbs for both humans and chickens along with some helpful growing tips. I based this list on what grows well for me in the chicken areas, what my chickens eat, and even where they rest during the day (in the oregano and lemon balm plants)—oh, and what I like, too! After trying these herbs, you'll definitely want to add more to your repertoire.

Rosemary (*Rosmarinus officinalis*): Rosemary is an evergreen perennial herb that's a member of the mint family and is native to the Mediterranean.

Hanging herbs in the coop can be decorative and useful.

Its ancient Latin name means "sea-dew" since it can be found growing close to the ocean where rosemary's blooms look like dew from a distance. In the United States it's grown as a perennial in warmer climes and treated as an annual or moved indoors into containers before the first frost in northern climes.

Rosemary has leathery, needle-shaped leaves and blue to purplish-pink flowers. It's a kitchen herb staple that goes well with Italian and Mediterranean dishes and it can be used in herbal breads, vinegars, oils, and butters. Rosemary makes a fragrant addition to potpourris and its essential oil can be used in bath products such as soaps and hair rinses.

In the garden, rosemary plants should be planted in full sun in an area with good soil and good drainage. Rosemary can be difficult to grow from seed, so it's better to buy it from the nursery as a plant.

In the coop, rosemary is a rock star as an insect repellent. You can grow rosemary near your coop to keep insects at bay. It can also be hung in bundles and made into a repellent spray by mixing rosemary essential oil with water in a spray bottle. Spray it liberally around the nest boxes, on the roosting bars, and near the doors and windows. You'll immediately notice a reduction in flies and bees that like to swarm in the coop. It even helps to keep away mites that like to burrow in the roosts during the day.

Rosemary also has pain-relieving properties and can aid in respiratory health. If it grows near your coop, your chickens can eat it at will. If not, you can harvest the leaves and sprinkle them in your chicken feed, around the coop, and in the nest boxes.

Lavender (Lavandula species): No herb garden or chicken coop is complete without lavender. This popular perennial herb hails from the Mediterranean and can be planted directly in the garden or grown in pots. It likes full sun and moderately fertile soil, but it can tolerate partial shade. There are many varieties of lavender to explore, from true lavender such as 'Munstead' or 'Hidcote' to wooly lavender varieties. It's important to read the growing information before you pick your plants. For humans, when we think of lavender, we think of bath products and relaxation. In fact, the Latin word *lava* means "to wash."

There are over 100 different cultivars of lavender. In most cases, it has spiky, grayish-green leaves and purple flowers. Both the flowers and leaves can be used dried or fresh. They can be used for seasoning meats, vegetables, desserts and drinks. They are also used decoratively in potpourris, wreaths, and sachets.

In the coop, lavender's calming properties make it ideal for sprinkling in nest boxes. If the chickens stay calm, they lay better. It is also an insect-repelling herb, so make sure to sprinkle the leaves and flowers around the coop and use them dried in the nest boxes, especially for broody hens that are sitting long-term. Lavender is a much better option for keeping pests away from broody hens than harsh insecticides that can be harmful to your hen and her young chicks. If you have room, lavender can be planted around the coop. Lavender is also good for circulatory health; it aids in digestion, helps heal wounds and insect bites, and relieves pain.

Thyme (Thymus vulgaris): Thyme is a popular, low-growing herb that's often used for groundcover or edging garden beds. It likes full sun with well-drained soil but will also easily adapt to partial sun. Thyme doesn't like to have wet feet, so go easy on the water.

Hanging herbs in the chicken coop, particularly around the nest boxes, can keep pests at bay and give chickens something to nibble.

There are lots of different cultivars of thyme. You can easily find the more common English, creeping, and lemon varieties. Then there are fancy varieties like lime, coconut, and lavender thyme. It's hard to stop with just one plant.

Thyme has strong antiseptic properties, can boost the immune system, is great for respiratory health, aids digestion, is an anti-parasitic, and can repel insects. It's great to sprinkle it around the coop, in the nest boxes, and in your chicken feed. It can be used fresh or dried.

Oregano (*Origanum vulgare*): Oregano is a Mediterranean herb that loves sun, easily tolerates poor soil, and is drought tolerant. I find it adapts easily to partial shade as long as the soil drains well. In Greek, the name oregano comes from two words, *oros*, which means "mountain," and *ganos*, which means "joy and beauty." So oregano literally means "joy of the mountain."

While oregano comes in handy on pizza night or for dipping with homemade bread, it's also handy for human and chicken health. It is said to be one of the most powerful antibiotics ever. Recently, a large commercial chicken farm started using oregano and cinnamon instead of antibiotics in their flock and cinnamon is now included in some nationally sold oregano-based supplements for poultry and livestock.

I chose to divide my original oregano plants and moved some of the divisions to the backyard where my chickens hang out. After it took, I let the chickens have access and they fell in love with my oregano. They spend long hours laying in it and nibbling its leaves. It's hardy enough that this doesn't bother the plant.

Since oregano grows easily, you'll likely have some extra, so go ahead and liberally sprinkle it throughout your coop. You can also add the dried flakes to your chicken's feed and even to the feed for your chicks in the brooder. Oregano is a great immune booster and may help prevent coccidiosis, blackhead, and even infectious bronchitis. Oregano improves respiratory health and aids digestion. Marjoram, which is closely related to oregano, can be used as a laying stimulant.

Mint (*Mentha* species): Mint is arguably the easiest of all the herbs to grow. Caution should be used when growing this perennial herb because mint spreads rapidly and can quickly take over a garden, so grow it where space is not a concern, such as on hillsides, to hold the soil in place. Or grow it in a pot to keep it contained.

Mint comes in many cultivars so it's easy to find one you like. There are the old standards of peppermint and spearmint. There are also fascinating cultivars with aromas like chocolate, orange, coconut, lime, apple, ginger, and lemon, just to name a few.

Mint can be used fresh and dried. It is usually used to add flavor to teas and other drinks. In fact, there's a specific mojito mint grown just for adult beverages! For the main course, mint pairs well with meats and fish. It's also used in desserts, jellies, salads, and as garnish.

Around the coop, mint is indispensible. It's a natural insect and rodent repellent. It has calming properties so it can help hens lay better. Since you'll likely have tons of mint on hand, it can be spread liberally everywhere, including the floors and nest boxes. You can also add mint essential oil to water and spray the nest boxes, roosts, and doorways. Mint also has naturally cooling properties, so it's a nice addition to the chicken coop during the summer.

Lemon Balm (*Melissa officinalis*): As a member of the mint family, lemon balm falls under the category of an easy-to-grow herb that can self-seed and spread rapidly if not kept under control. Mine has spread far and wide to various parts of my front and back yards, including my chicken areas.

Lemon balm can be minced and used fresh in salads, sauces, and sprinkled over vegetables, meat, and fish dishes. It's commonly used in teas and potpourris.

More importantly, lemon balm is associated with stress reduction and calming. It also has antibacterial properties and is a natural rodent and insect repellent. You can sprinkle lemon balm around the coop and in the nest boxes. My chickens love lemon balm and can often be seen taking a nip or two of the leaves as they walk by it.

Other Useful Plants

BASIL
Improves general
  good health
Immune system booster
Improves respiratory health
Improves circulatory health
Improves bone health
Repels insects
Reduces stress
Provides orange egg yolk

CATMINT
Natural wormer
Natural insect repellent

CHAMOMILE
Aids in relaxation
Improves digestive health

CILANTRO
Improves general
  good health
Antioxidant
High in vitamin A and
  vitamin K

DANDELION
High in calcium
Improves digestion
Improves kidney health
Improves liver health
Increases bone health
Relieves pain

DILL
Aids respiratory health
Improves digestive health
Reduces stress
Relieves diarrhea

FENNEL
Egg-laying stimulant
Improves digestive health
Promotes general
  good health

LEMON VERBENA
Increases relaxation
Improves respiratory health

MARIGOLD
Provides orange egg yolk
Repels external parasites and
  insects
Natural insecticide
Aids in respiratory health
Anti-fungal
Relieves diarrhea
Improves liver health
Heals wounds and
  insect bites

NASTURTIUM
Natural wormer
Laying stimulant
Helps to heal wounds
Improves digestion

Relieves congestion
Repels insects

PARSLEY
Egg-laying stimulant
High in vitamins
Promotes circulatory health
Promotes bone health
Promotes good digestion

PINEAPPLE SAGE
General health

ROSE PETALS
Relaxation
Antibacterial

SAGE
Immune system booster
Improves overall health
Antiseptic properties
Prevents internal parasites
Combats salmonella and
  Escherichia coli
Repels insects

YARROW
Antibacterial
Increases relaxation
Insect repellent

Summer

We look forward to summer with its pool parties, cookouts, and camping. Long summer days and nights are the stuff of magic and memories. For our chickens, summer is arguably their toughest season because heat can be harder on them than cold. Think about it: there are ways to get warm in winter, but in heat, it's tough to get cool. This is especially true when you start out with a higher body temperature, like chickens.

During summer heat, chickens are susceptible to the same heat-related complications as humans including heat stress, heat stroke, and even death. I once had a Barred Plymouth Rock that was relaxing in the deep shade of a tree line with her flock mates during an extremely hot and humid summer day. I brought everyone some fresh water. She drank heavily and then literally turned around and dropped to the ground dead. Her death took place so quickly that it took me a minute to process what had happened. And while I'll never know for sure whether she had

Chickens don't sweat, so panting is often a first line of defense when dealing with hot weather.

some type of underlying condition, it's certain the heat was the overriding factor that caused her death. So while you can't stop the heat and all the emergencies that arise because of it, luckily there are strategies to help your chickens through the summer.

Chicken Behavior and Adaptations

Chickens automatically know what to do to keep themselves as cool as possible. Chickens do not sweat though, so it's harder for them to cool down than humans. Laying hens are more prone to heat-related issues since their bodies are working hard to produce eggs. As the temperature climbs, a chicken will start to pant. For fully grown birds, this happens at around 85° Fahrenheit. For chicks, it's around 100° Fahrenheit. You'll see your birds open their beaks and begin to breathe a little harder. This allows hot air to evaporate from the lungs. You'll also see them hunker down with their wings held partly open away from their body. This increases the surface area that can be cooled.

Note

Be aware of not only the air temperature but also the humidity. The higher the humidity, the worse the heat is for chickens. Humidity decreases the amount of evaporation a chicken can achieve through panting.

During hot weather, you'll notice your birds will not move around a lot, especially during the hottest parts of the day. This behavior is much like human behavior in hot weather and can help the body conserve energy and stay cooler.

Many chicken breeds have adaptations that help them better regulate their heat. You'll notice that Mediterranean breeds such as Leghorns, Minorcas, and Anconas have extra large combs. They give the chickens a comical look as their combs flop in odd ways while they're going about their day. But in the heat, they serve more than a comedic purpose. They serve as air conditioners allowing heat to radiate from the body. As blood travels through the comb it is cooled and that cool blood travels back through the body.

For all chickens, you may notice their droppings during hot weather are much more watery. This is not an indication of sickness; rather it's an adaptation to help cool the body by expelling moisture.

Heat stress and exhaustion are the common side effects of excess summer heat. They are cumulative, so make sure you keep an eye on your birds closely during heat spells. While chickens pant at 85° Fahrenheit (29°C), heat-related problems can start with temperatures as low as 80° Fahrenheit (26°C)—especially if there's high humidity.

The first sign of heat stress is that your chickens are eating less and drinking more. You'll also start to notice a lack of egg production. More serious signs are labored and rapid breathing, excessive drinking, weakness, unsteady walking, or lying prostrate on the ground with eyes closed. Do not panic and bring a bird that's suffering major symptoms into the air conditioning. This can cause shock. Instead, try to slowly decrease the bird's temperature by moving it to shade and dunking its feet in cool (but not freezing) water. You can also dunk the bird up to its neck if needed. Keep your chickens cool and hydrated.

HEAT-TOLERANT CHICKEN BREEDS

If you live in a region with extended and high heat, it's a good idea to keep a flock of heat-tolerant birds. Here is a list of some popular breeds that fit the bill.

- Ancona
- Barnevelder
- Barred Plymouth Rock
- Easter Egger
- Leghorn
- Minorca
- Rhode Island Red
- Sicilian Buttercup
- Sussex
- Welsummer

Hot Weather Chicken Keeping

Knowing how chickens behave and adapt to hot weather helps to guide our actions as chicken-keepers during summer.

Keep the Water Coming: We know chickens increase their water intake in the summer, so it's important not to skimp. Don't make your chickens work for their water. If you notice your chickens like a particular cool spot in your yard, set an extra watering station up there. In fact, it's good to have water set up in various favorite spots. Just make

Brown Leghorns are a Mediterranean breed that tolerates heat well.

sure to keep your waterers out of the sun as that can quickly heat up your water and do more harm than good for your birds.

You can also add electrolytes to your chicken's water to replace minerals and nutrition lost in hot weather. There are poultry electrolyte replacements that you can buy online or at your local farm supply store. They come in small packets that almost look like yeast packets, and are meant to be mixed with water according to the directions.

Provide Lots of Shade: Shade is important for obvious reasons, but even more so for birds that are dark colored, such as Black Australorps. Their

Watermelon makes a great summer treat. Bonus points for standing on the melon and eating at the same time!

feathers don't reflect sunlight very well and instead absorb the heat. So they can get overheated much faster than their white and light-colored counterparts. If a bird has no shade outdoors, it will tend to stay indoors. This can cause overheating since there's usually very little air movement indoors. Consider adding a fan to your coop if your birds are confined. Free-range chickens will automatically seek out the coolest places in their environment. If your yard doesn't have much shade, then it's a good idea to add it through your plantings or through built structures such as decks and patio tables. Whatever gets the job done!

Ventilation: Whether your birds are free range or confined, they all have to go back in the coop at some time. Make sure that all the time they spent outdoors trying to stay cool isn't wasted when they go into a coop that's as hot as a sauna. Try adding lots of screened openings to let cool air inside.

Outfit a screen door with predator-proof hardware cloth and use that instead of your usual coop door. Keep all the windows open. Use roofing ventilation to let the hot air escape as it rises. There are multiple roofing ventilators at the hardware store that don't require a fan. When properly installed, they keep the elements out and allow heat to leave the coop.

Keep Activity to a Minimum: Your birds already know to do this. They'll stay very still on a hot day. So don't go outside and place treats or water far away or decide it's time to cuddle and play. Let your birds stay still and avoid disturbing them especially during peak heat hours.

Cool Things Down: We love cool and refreshing summertime treats and your birds will too. Refrigerate a watermelon or other melon, such as honeydew, cut it in half, and watch your birds devour it. Not only does this provide some coolness, it also provides extra water intake. Freeze berries in water and then have fun watching your birds peck at the cool ice to get the treats inside. Set up a mister (not so much that it forms puddles) and let your birds decide to take a cool stroll through it. Or set up a baby pool with some water (not too much) and let your birds dip their feet.

Cut Back on the Grains: Scratch grains and whole dried corn kernels take a lot of energy to digest and this raises the chicken's body temperature. You can feed scratch grains and corn in the summer but it's a good idea not to overdo it. As with any treat, keep it to a maximum of 10 percent of your flock's diet. And be careful when you feed scratch grains during the summer. Many chicken keepers will eliminate scratch grains altogether or feed them only during the cool morning hours or on cooler summer days. This is not a good treat during summer heat waves!

Fall

I find fall to be a welcome time of year after the stress of summer's heat. It's the perfect time to check on your coop to make sure it's well fortified for winter before bad weather comes. The break in temperatures means less worry about heat-related conditions. Fall can be finicky, like spring, so make sure to watch for heat spells. Fall is also a time where hatcheries are winding down and starting to give some customer bargains. Just remember that if you buy chicks during this time, they will mature into laying hens during the winter. You won't get eggs from them until spring. During fall's respite from harsh weather, your birds will molt. This is an important part of their life cycle that can be assisted through some easy chicken-keeping methods.

Molting

Molting is the natural process in which chickens shed old feathers and replace them with new feathers. Many people think of molting and older chickens, but it actually happens at different times throughout a chicken's life.

Molting starts when your chickens are young. Between 1 and 5 weeks of age, baby chicks will gradually lose their down feathers for their first

This young Easter Egger is undergoing her final molt to get her first set of adult feathers.

set of juvenile feathers. At 7–9 weeks of age, juveniles will replace many of their baby feathers with a full set arriving around 12–13 weeks of age.

After 10–14 months of egg production, hens will molt. Roosters molt too. Taking into account their chick phase, molting usually happens around 18 months of age and each year afterward. It can take place in spring, but usually takes place in late summer or early fall. This timing is especially important because feathers equal warmth. Going into the late fall and winter with nice, new feathers means you'll be warmer during the cold months.

Molting is typically triggered by decreasing daylight hours, which is why it's most common in late summer and early fall. It can also be caused by stress, lack of proper nutrition, lack of water, and hatching a clutch of eggs.

Molts come in two types. A soft molt is a gradual process that may not even be noticed as individual feathers fall out and are replaced. A hard molt is impossible not to notice as a chicken loses most of its feathers almost at once. A full hard molt can be a scary site with a nearly naked chicken and usually lots of pin feathers, also called blood feathers because they are still connected to a blood supply, poking out like a porcupine's quills. Pin feathers are covered with a hard waxy coating that falls off as the feathers come in fully. It takes 3 weeks for a feather to be replaced, with the whole molting process taking about 8 to 12 weeks to complete.

Many people think their chickens are sick during this time. Molting chickens are not sick but this process is hard on them. You can tell a lot about a chicken's productivity by its molting process. Hens that are good layers will usually molt hard and fast. They can be irritable and

During molt, exposed pin feathers look awkward and can be painful to the touch, but they will result in beautiful new feathers.

dull-looking before molt. Hens that aren't great layers will typically only have a soft molt. You may never even notice they molted.

Molting chickens lose and re-grow their feathers in the same sequence. First they lose their head and neck feathers, then the feathers on their back, breast, and thighs. The last feathers to go are their tail feathers.

During a molt, hens will decrease or stop laying eggs altogether and will instead spend this time building up their nutrient reserves. They also have to make new feathers during a molt. Feathers are made of about 85 percent protein. This protein is called keratin and it's found in many different animals throughout the world, from our fingernails to rhino horns. So be sure to add extra protein to your flock's diet while your birds are molting. A good way to add protein is through high-protein snacks like sunflower seeds, nuts, peas, meat, mealworms, and soybeans. There are also higher-protein commercial feeds that can be fed during a molt.

Fun Fact

Eggs laid after molting is complete are usually larger with improved shell quality. Typically production drops about 10 percent after a molt.

As your adult or juvenile birds are molting, remember not to handle them too much. Growing feathers can be painful to the chicken when touched. Also, make sure to watch for feather picking or scratched skin during this time. Since a chicken's skin is exposed, bloody areas can attract unnecessary attention.

Small combs and wattles are helpful during cold weather because they are less susceptible to frostbite.

Winter

Winter is the other extreme weather season that makes chicken keepers cringe. No longer is it pleasant to pop outside and give the chickens water and maybe linger as you watch them free range or stop to give them a treat. Eggs have to be gathered often before they freeze. You need to bundle up before making the trek to the coop. In general, chicken keeping gets a little harder.

Chicken Behavior and Adaptations

While summer's heat is extra tough on chickens, the cold is a little easier on them. Chickens have adaptations to help them through the cold and when combined with good chicken-keeping methods; winter doesn't have to be a problem.

Just as in summer, chickens know what to do when cold hits. You'll see them stay indoors more. They will roost more throughout the day and they'll roost closer together. They'll also find spots in the chicken yard to gather closer together since huddling is a great way to share body heat. They'll seek out shelter from wind and precipitation. If the sun is shining, they'll find that too. Chickens will puff their feathers, which creates spaces to keep warmer air close to their skin. Chickens can also be seen tucking their beaks under their wing feathers to breathe in the warmer air.

Chickens naturally run at a higher temperature and in the winter their metabolism speeds up to keep them warmer. Like all vertebrates, chickens have a closed circulatory system. Their system has special adaptations that help them conserve heat and protects their exposed legs and feet.

Chickens, like humans and vertebrates in general, have a four-chambered heart that separates oxygenated and deoxygenated blood. Arteries carry bright-red oxygenated blood away from the heart and veins carry darkened, less oxygenated blood back to the heart. These veins and arteries intertwine so they are close together in a specialized adaptation called *rete mirabile*, which means "wonderful net" in Latin. This wonderful net of blood vessels produces a countercurrent heat exchanger that's like having built-in central heating. As warm blood leaves the heart and goes to the extremities it warms the cooled blood returning from the extremities. This conserves heat from the body core. If you have a mixed flock with chickens and ducks, it's good to know that ducks also have rete mirabile.

Fun Fact

When birds stand on one leg and tuck the other into their feathers, it halves the amount of heat lost from their limbs.

The other exposed part of a chicken's body is its comb and wattle. It's funny because in summer's heat, you want the comb and wattle to be large to help cool the body. In winter, you want the comb and wattle to be small to minimize heat loss and the chance of frostbite.

COLD-HARDY CHICKEN BREEDS

If you live in a region with extended cold periods, it's a good idea to keep a flock of cold-hardy birds. Here is a list of some popular breeds that fit the bill. Most cold-hardy breeds are medium to large weight, 6 or more pounds, with smaller combs such as peacombs.

- Ancona
- Barnevelder
- Black Australorp
- Brahma
- Buckeye
- Buff Orpington
- Cochin
- Delaware
- Dominique
- Easter Egger
- Jersey Giant
- Marans
- New Hampshire
- Plymouth Rock
- Rhode Island Red
- Sussex
- Wyandotte

Cold-Weather Chicken Keeping

While warm-weather chicken keeping is pretty straightforward, cold weather brings out strong opinions among chicken keepers about heating and lighting. It's good to look at all the options for keeping your birds comfortable and safe, then use the methods that work best for you and your flock.

Ventilation: The number-one mistake that chicken keepers make while trying to keep their birds warm is to make the coop too tight. While it's true that drafts are bad for chickens, an airtight coop can be even more dangerous. The moisture from respiration in an airtight coop builds up, leading to wet air and an increased chance of frostbite. Plus, the ammonia from the chicken's droppings builds up and can damage their lungs. Moisture and ammonia need a place to escape.

Add the Grains: Chickens need to eat more in winter to keep up their warming metabolism. So make sure feed is always available. Adding scratch grains and whole dried corn to their feeding regimen raises a chicken's body temperature as they take a lot of energy to digest. You can also feed suet to your chickens as the fat provides warmth. Try feeding these warmth producers right before your chickens roost for the night. It will give them extra comfort through the coldest hours.

Speckled Sussex (above) and Buff Brahmas (below) are both cold-hardy and family-friendly chickens.

Q&A: Chicken Keeping through the Seasons

It's so cold outside in winter. Should I heat the coop?

If you've taken winter chicken-keeping precautions, then no extra heat is really necessary. Chickens in a dry, well-ventilated coop with fresh food and water tend to do fine. You'd be surprised how warm a coop can get just from the body heat of your flock. As humans, we notice that in the beginning of the winter it feels really cold, then our bodies gradually acclimate to the colder temperatures. As winter goes on, the cold bothers us less and less. This is the same with chickens, they gradually acclimate to the climate. If they have a heated coop, they can never really acclimate, and winter is much harder on them.

How cold is too cold for adult chickens?

This is a common question that goes along with "should I heat the coop?" especially when people think about the brutally cold winters in the far north. I have friends who raise chickens in these types of extreme climates and have no problem with their chickens even in sub-zero temperatures. I have seen statistics that say chickens really don't suffer until it gets to minus 20° Fahrenheit in the chicken coop. As mentioned on page 158, extreme heat is far worse for chickens than extreme cold. That's not to say that bad things can't happen to your chickens in cold weather: for instance, frostbite. There's no set temperature where it's too cold for chickens. In extreme weather it's important to check on your chickens often and be prepared to react if the need arises.

Is it a good idea to bring my birds into the air conditioning on hot summer days?

Although we humans love our air conditioning, it's not a good idea to bring a bird that's used to warmer outdoor temperatures inside just to make the bird more comfortable. Just as chickens acclimate gradually to the cold weather, they also acclimate to the warm weather. Have you ever noticed how hard it is, as a human, to go from air conditioning straight into the outdoor heat of summer? This is even more so for chickens. It's hard on their bodies to keep re-acclimating.

Can chickens molt at other times of year?

Yes, as not all chickens follow the written rules. If an errant molt takes place, it's important to check the health of your bird. It can be just a quirk from a chicken with bad timing. But there are other reasons for unscheduled molts. Stress can be a big factor, whether from lack of proper water or food or extreme heat. Improper lighting can also play a part. (This can happen when a coop is regularly lit during the night and then that light is removed.) Molting can happen after a broody hen hatches her eggs as well.

Applying a light coat of Vaseline or Green Goo can help prevent frostbite.

Add Bedding and Keep it Clean: Increased bedding can provide a nice insulator and cushion as birds spend more time in the coop through the winter. Some folks will actually use straw bales lined along the walls as insulators and places the chickens can perch. Just make sure that your bedding choice stays clean and dry through the wet and cold months. Wet bedding breeds disease and is especially hard on feather-legged and feather-footed birds.

Add a Snow-Free Zone: Even in winter, chickens benefit from fresh air and exercise. An enclosed run with a roof is ideal. If your run doesn't have a roof, you can add tarps or pieces of wood to the top to keep the snow out. You can wrap the run with tarps or plastic on the windy sides to keep out harsh winds and flying snow. If light snow is predicted, you can lay down pieces of wood or old baby pools upside down in the yard before the snow. After the snow, remove them for snow-free areas. If snow is deep, shoveling some pathways to sheltered areas, like decks, allows your chickens to safely roam.

Don't Let the Water Freeze: Chickens need water to live and ice just won't cut it. You can change the water often throughout the day or you can purchase a heated waterer to make the job easier than hauling water all day, but one way or another that water has to stay free of ice. It's a nice treat on cold winter days to bring your flock warm water. It will help keep their bodies warm without using body energy to make the warmth.

Keep Combs and Wattles Protected: If you've got roosters or birds with larger combs and wattles, it's a good idea to cover them with petroleum jelly for added protection and frostbite prevention. Just take the petroleum jelly and generously massage it into the red, fleshy areas leaving a fine covering of jelly. Be sure to regularly check the

Chickens benefit from getting out of the coop and taking in some fresh air, even when it snows.

combs and wattles and apply more petroleum jelly if the area feels dry. If temperatures go below freezing for an extended period of time, petroleum can freeze and become useless so it's best not to rely on petroleum jelly alone. Also, try to raise the height of watering areas for your roosters. Those long wattles can hang in water if it's too low and then easily get frostbitten.

Light the Coop for Egg Production: As we know, light stimulates a chicken's laying hormones, so the lack of daylight hours in winter means egg production drops. To light or not to light is a personal choice. Some folks let their hens take a break from egg laying during the winter. Others choose to light the coop to keep egg production consistent. It doesn't take much light to do the trick. The two most important things to consider when lighting a coop are fire safety and timing. Fire safety speaks for itself; make sure your lighting is securely hung and not near anything flammable. With timing, remember that chickens require 14 to 16 hours a day for optimum egg production. Count the number of winter hours of daylight you get and then add the extra hours needed. Add them in the morning and not at night so chickens can get adequate rest. Plus, it's hard if you're lighting in the evening and then turn out the lights all at once. Chickens can't see well in the dark and this will leave your chickens unsettled and not roosted for the night. Consider using a timer so you can get adequate rest too! And feel free to get festive with the lighting, like using Christmas lights. They're pretty, they give enough light to stimulate laying, and they're less of a fire hazard.

CHAPTER 9

Coop Truth

To understand chicken coops at my house, you have to understand the space we have for our chickens. Although we own almost 13 acres, our acreage is hilly. I'm not sure we own any level land that actually started out that way. To make things harder, our backyard can only be accessed through a gate that's bordered on one side by our pool and the other side by our garage. The good thing is that we do have a nice-sized fenced backyard where our chickens roam freely each day and other free-range areas. But getting our feathered flock a proper coop was not an easy task.

Yes, we have our share of local stores that sell beautiful pre-made coops that you can easily have hauled to your home and set on the appropriate site. I found a few of those that I really liked and that would have been my first option. However, none of them would fit through my gate.

There are companies that sell nice coops that come unassembled and just require a few hours to put them together. But in a delusion of grandeur my husband declared he could build us a coop; problem solved.

I reluctantly agreed to this plan, not once but twice. I am proud to say that both coops are still standing and housing our birds quite nicely. But getting to the finished product wasn't always easy. My husband and I had our share of disagreements along the way. He'd want a window here; I'd want a window there. I'd ask for more ventilation; he'd say we had enough.

The biggest disagreement we had came when we built our second coop. The need for this coop came since we had three roosters in our flock. The two brothers would occasionally fight and this was scary and resulted in too many wounds. Worse than that, one of the brothers, Nate, began attacking our hen, Little Muff. We couldn't get him to stop; he just had it out for Little Muff. But we weren't willing to kill him either. He got along well with our oldest rooster, Roopert. So we decided to make our old coop a bachelor pad and move the rest of the flock to some new digs.

We got started on construction and things were going well. My husband dug the footers and leveled them off. He had the floor of the coop assembled. And then, he asked me to take a look; a bad idea on his part. I walked around the coop. I walked on the floor itself. I laid down on that floor. And, I just couldn't shake the feeling that it was too small. I thought another couple feet would be best and give us room for any plans that may come along. That would take the coop from 8×8 to 8×10 feet.

You can image how well received my suggestion was! But my husband agreed to more footers and an expansion of the coop size. It took some extra time, extra work, and a few extra trips to the hardware store, but in the end, even my husband will admit it was a good adjustment to make. Now we can walk into our coop and have plenty of space to move around. The coop isn't too big if our flock numbers are low and isn't too small when we want to add more chickens.

It's funny because when you're starting out, the focus is all on the birds and how to raise and care for them. But don't underestimate the value of a good coop and what it may take to get your birds properly housed. The coop is often addressed between the time you get your chicks and the time they're ready to begin life outdoors.

The reality is that's not a huge time frame and there may be some compromises that have to be made along the way. Once you've got the coop and your birds are set up, you may find there are some tweaks that need to be made to make life easier for you and your chickens. Don't be afraid to change the coop; it's kind of like a garden, ever changing and evolving through the years.

Lessons from the Coop

Besides choosing and getting your chickens, the biggest decision you make as a beginning chicken keeper is what coop to use. Before you bought your first flock, you probably thought one coop would do it. But I have to say, even with the best of intentions you'll probably need more than one coop—or at least different coop variations as you move forward and expand your flock. Even without expanding, there are some mini-coops you may want as your flock matures.

After years of raising hens and roosters and various versions of coops along the way, here are some lessons I've learned.

Carefully Consider Coop Placement

In the beginning, everyone is caught up in how much square footage you need per bird, roosting bar length, and proper quantities of nest boxes. In fact, the list of coop requirements seems endless for the beginning chicken keeper. Along with all those requirements, you often get the advice to

locate the coop as far away from your house as possible because it smells. My advice is to think long and hard about this.

We located our first coop, which is still standing, as far away from our house as possible. Since we live on a hillside, the coop had to be set on a flat spot that requires a trek across the yard and up a hill. That coop has many advantages because it's huge, shaded from hot sun, and protected from really bad weather because of the rise it's on. But its disadvantages became its doom and the reason we only use it as a bachelor pad today. Carrying water up a hill several times a day gets tough. Carrying any other supplies, such as food and bedding, also gets tough. All the traffic eventually kills the grass. When it rains or is snowy and icy, then heavy loads become treacherous and dangerous. Also, what if you want to run electricity to the coop? It's so far away that it takes tons of extension cords. Then when friends and family come to babysit your chickens while you're away, they also suffer through all the same inconveniences.

We eventually built a new coop at the bottom of our hill, close enough to easily run a water hose or an extension cord. It's close enough that nothing heavy requires great physical exertion.

Does it smell? No. We'll discuss some cleaning strategies later in this chapter. But for now, sufficed to say, we clean our coop often and it doesn't smell. We painted and decorated our new coop so it's a backyard attraction. We did have to give up our ideal front door placement that faced out of the weather, but that was a small price to pay and we've learned to work around that. Plus, now when we go out of town, no one minds watching the chickens for us.

This grow-out coop is moveable and allows the young chickens to get outdoors while the older chickens can also see and inspect their new flockmates.

Plan for Grow-Out Options

You've got your new chicks and they're growing up fast. This can be a little sentimental, but that sentimentality quickly turns to action. With the first flock, those chicks just move out to the coop when they're old enough and there are no worries. However, unless your coop is big enough to accommodate everything and everyone, then there's a time as you're expanding when your chicks are not babies anymore but they're not big enough to be integrated into the adult flock. They can't stay in your house or garage all day. They need a safe place to get outside and roam and learn to be chickens. So what do you do?

For this, there is no right or wrong answer, only what works best for you and your chickens. I've been around enough chicken keepers to see many variations at this juncture. Some folks will partition off part of the existing coop and house

Chicken History

I was walking through an antique show when I saw an interesting tin chicken coop. Of course, I didn't buy it right away. This was my first time at this show and I wanted to look around a bit. But that coop stayed with me. So, as I was getting ready to leave, I popped back over to that booth and the coop was still there. It was meant to be mine! But what was it? I could see a rooster logo on the front with the name "Cruso" scrolling through it. This was obviously a company of some type, but I had never heard of it.

I asked the seller if he knew any history behind this piece. He thought it might be a transporting crate for chickens, just shaped to look like a coop. He also surmised it could be a cage used to carry roosters to a cock fight. I looked closer. I didn't say it, but I knew none of these assertions were true. For one, the coop was too big to easily handle. (It's about 2 ½ feet tall and about as long and wide.) It was also extremely heavy; too heavy to carry long distances. Plus, the detailing was too well done for what the seller thought was essentially a transportation box like a cat or dog carrier.

Even not knowing exactly what I had found, the price was right and the coop would look great as a decorative piece in my house, so I bought it. And then I researched it the moment I got home.

Usually with pieces with logos, you can find another of its kind and get some sense of history. With this, I couldn't find a thing. I looked everywhere, even on decorating and craft websites. Surely someone else was decorating their home with this; chicken-inspired décor is hot.

Then I ran across someone who posted a coop that looked similar to mine in size but it had a different company logo and some different detailing. Then I found someone else that had a rounded, open-air coop that looked like a chicken run. Same size, same metal, just a different logo.

It turns out that I had come across a salesman's sample coop. In days gone by, there was no Internet or easy way to advertise and sell to your customers. You couldn't just scroll a company

the new chickens right there. Many choose to do this from day one. (This is where placement of the coop comes in handy, because unless you're using a broody you'll need electricity to keep the babies warm.)

I like to use what I think of as a stepping-stone system. I brood inside the house, then move the chicks to the garage as they don't need heat anymore and become too messy to stay inside. I have an outdoor chicken run that's safe for the chicks to have time outside, but I move them back into the garage at night or when I'm gone and can't supervise them. This allows them safe free-range time where they can see the adult flock members, but keeps them safe at night.

Another option is to use a grow-out coop. This is usually a smaller coop with an attached run; in some cases, it's the first coop from a flock that has now outgrown their digs. These smaller coops are usually readily available at farm stores. They're

This antique salesman's coop could be taken to a potential client so they could see all the features of the coop they were considering.

website and look at all the beautiful options for sale and then have it delivered.

But, people still needed coops. Chickens were a farmer's staple. So salesmen would come to your farm and have some sample coops on hand for you to shop. That's why the detailing on mine is so accurate, complete with vent holes and a working door.

The key to figuring this out was actually the logo on the posting of the other sample coop I saw. It said "Hibbard" on it. I researched that company and found out that referred to Hibbard, Spencer, Bartlett & Company; a hardware retailer that introduced the Cruso brand in 1902. The Cruso brand sold many products with that same rooster logo on them, from padlocks to waffle makers. It turns out I had found an early 20th century piece of chicken keeping! And after looking at the coop, it's clear that not much has changed in the world of chicken coops since then.

great for getting the chicks outside while still keeping them separate. They also make a great coop for a broody and her chicks, a sick bay, or isolation coop. Just make sure it's predator-proof so everyone stays safe.

There are many DIY options when it comes to chicken coops and chicken keeping. But be realistic about your carpentry skills and ability to get things done. If your carpentry skills aren't up to snuff or your other obligations don't allow

you much time to spend building things, then consider buying pre-made options or getting creative. Do you have an old garden shed that's not being used? If so, it can be easily transformed into a wonderful chicken coop. Do you have neighbors and friends who might exchange a helping hand for delicious farm-fresh eggs later? Don't let your chicken coop become a source of frustration. Make it easy and fun and you'll be more likely to stick with it for the long haul.

Make Sure Your Coop Fits Your Needs

For me, family vacation and getaway time is important. So everything in my coop is oriented toward everyday ease and convenience as well as the convenience of potential caretakers. Even if vacation is only once a year, I want to know that I can get away and my chickens will be safe and comfortable. But vacation wasn't my only thought when designing my coop. I work from home, so my chickens free range in fenced areas during the day (I do check on them and visit often) and put themselves away at night. After they've perched and settled in, I shut the coop door so they can rest safely. This works well when I'm around all day though it may be different if you work outside the home. Regardless of work schedules, what about nights out for dinner and a movie? What about times when you may want to take a day trip and won't be back before dark? How about nights when you're at sporting events for the kids or other school activities? Those are times you may not be able to get home to do a final check and shut the door after your chickens have roosted. For those times, in addition to a large coop, I've got an attached fully-enclosed run. It's predator safe. That way my chickens have room to safely roam outdoors and indoors, and I don't have to worry about getting back at a certain time. On days when I'm going to be gone, I'll let them stay in their coop and run instead of free ranging. If I'm just leaving to go out for the night, I'll round everyone up into the coop and run before I leave. That way they're safe, happy, and have access to fresh air and outdoors. So, as your chicken keeping progresses, it's good to take a look at your coop and see if it's meeting your needs and make adjustments accordingly. You'll be thankful you did.

Inside the Ideal Coop

Chicken coops these days come in all shapes and sizes. There are coops that cost thousands of dollars and are designer in every way. There are adorable backyard coops meant to grace the lawns of suburban backyard chicken owners. There are converted garden sheds. And there are creative coops, made out of pallets and recycled materials.

No matter what coop you choose, to be honest, the chickens really don't care what it looks like on the outside. They're much more concerned about whether everything they need to roost at night, get out of the elements, and lay eggs is inside. The ideal coop for chickens is much more utilitarian.

Space Requirements

Starting out, most people figure out a general space requirement and go from there. But as you expand your flock and possibly add birds of different sizes, such as bantams and heavies, you need to consider the varying size minimums. Remember guidelines are just guidelines and most people, including me, believe that more room is better, up to a point. However, sometimes too big is just too big! This comes into play during the winter. Body heat equals warmth, so too few bodies in too big a space won't help the flock stay warm. For instance, 10 bantams in a 10 foot × 12 foot coop can't keep it warm. You've got a lot of dead air space in there!

If you've inherited too big of a space, there are ways to cut it down so your chickens are comfortable. For instance, wall off a walk-in area where you can keep food, extra bedding, and supplies. You can also lower the ceiling by adding an elevated loft storage space. Both of these options are great because they limit the space your birds have to keep warm and they add much-needed storage. And really, too much storage is never a problem!

INTERIOR COOP SPACE REQUIREMENTS

Heavy Breeds	4.0 square feet per bird
Standard Breeds	3.0 to 3.5 square feet per bird
Bantam Breeds	2.0 square feet per bird
Exterior Run Space Requirements	8.0 to 10 square feet per chicken

Roosting Bars

In simple terms, the roost is where your birds will sleep at night. It mimics the tree branches where they would perch to stay away from predators if

These roost bars are made of sturdy tree branches and spaced so the chickens can hop on and off easily.

Nest boxes are popular spots and the favorites are highly sought after. They should be kept clean with absorbent bedding for clean, unbroken eggs.

they were living in the wild. You should provide nine to 12 inches of roost space per bird.

Roost bars are usually staggered, since a backyard coop most likely doesn't have enough space to give each bird a foot of space all at the same height. Roost bars can be made from 2×2 or 2×4 lumber with the edges rounded or from sturdy tree branches. They should not be made of plastic as it's too slippery for chickens to get a good grip. Likewise, they shouldn't be made of metal since that can freeze in winter and cause frostbite. With 2×4 pre-cut lumber, there is debate about using the 2-inch side for the birds to perch or laying

it flat with the 4-inch section for the birds to perch. If you have the ability, it's nice to mix up your perches and let your birds decide what they prefer. The bars can be laid out like a tilted stepladder with each rung being separated by about a foot and no more than 18 inches apart. This will ensure the birds have an easy way to get on and off the perch without harming themselves—especially if you have heavy breeds like Brahmas and Jersey Giants.

It's important not to place nest boxes or water and food below the roost bars since chickens will defecate throughout the night. Many chicken

A landing bar in front of the nest box helps to reduce injuries
as hens and roosters hop in and out.

keepers will add a droppings board below the roost bars so they can easily remove it for daily cleaning.

Nest Boxes

The 101 rules for nest boxes say to have one box for every four to five chickens. They should be 1 foot wide, 1 foot tall, and 1 foot deep. Try not to mount them on a north-facing wall so they stay a little warmer in winter. Keep them in a private spot away from excess light and traffic, add a landing board or roost in the front of the boxes so hens can fly up and have a place to land, and consider adding nest boxes on the outside of your coop for easy egg removal.

This is all great advice and should be followed. But in the end, your chickens didn't write the 101 rules and probably won't always follow them. You may have perfectly good nest boxes all lined up and all the same, but your chickens have their preferences. Even though their preference can change from day to day, you may end up with all your hens vying for just one box. This can create quite a bit of commotion. I have one Brown Leghorn that stands below the popular nest box as it's being occupied by another hen and vocalizes loudly and continually while stomping back and forth. Other hens are more proactive and use the "pile on" approach, trying to fit as many bodies into one nest box as possible. You may go to the coop and see multiple hens in the same box all looking a bit grumpy. These antics are fun to watch and make great stories, but they can lead to broken eggs and a general mess. Some chicken owners swear by adding fake eggs to the other nests to make them seem better. But in the end, even chickens succumb to the allure of what's popular and want to be in the same box as their compatriots.

As you add chickens to your existing flock, there is no need to worry about the basic rules of showing your hens the nest boxes and enticing them to lay in that spot by adding fake eggs. Your existing flock will lead the newcomers by example, and they will follow suit. If you find eggs that have been laid on the floor of your coop or in other places, it's a good idea not to eat them. They have likely not been lying in the most sanitary conditions, so it's best to air on the side of caution.

Many people wonder about nest box curtains that cover the box by hanging in front of it. Are they just a frivolous way to decorate the coop or do they serve a more utilitarian purpose? The answer is no and yes. They do make great coop decorations, but they also serve a good purpose. Here's how nest box curtains can be helpful:

- They reduce extra light and give more privacy.

- They can encourage a broody hen by giving her privacy, and they can discourage others from becoming broody by blocking their view of her.

- They can help discourage egg eating since it's harder to see all those yummy eggs.

- They can add warmth in the winter making the nest boxes more comfortable for laying and keeping eggs from freezing.

- If you shut the curtains at night, they can discourage sleeping in the nest boxes.

- They can discourage vent picking, which sometimes happens as other hens see the red swollen vent of a hen that is laying and start to pick it, causing damage.

Dust Baths

We know that dust bathing is an essential chicken behavior to remove old skin and oils and to keep feathers healthy. If your chickens free range, they will likely pick their own dust-bathing site, like the base of a tree facing the sun, a sheltered spot under some shade, or in your favorite plant bed. That's perfectly acceptable and you don't have to make a dust bath for them. If your chickens don't free range, it's good to make your chickens a dust bath with some proper materials. You may find even with the free rangers, a homemade dust bath can be the ticket to keeping them from digging their own baths all over your yard.

Dust baths don't have to be complicated; in fact, they're kind of like building a big sand box. A 4-foot × 4-foot box is easy if you cut two 8-foot 2×6 pieces of lumber in half and turn them into a square. If you have a small flock, you can get creative with the structure. I've seen people use concrete mixing pans from the hardware store, large flower pots, and even a baking pan. Just make sure it's deep enough and wide enough for your chickens to fit in and dig. This will depend on the size of your birds. Young chickens and bantams need less space to spread out and bathe than standard and heavy breeds. Once you've got the structure, then fill it with dirt and sand as the base. You can add diatomaceous earth for its external pest-repelling properties but add it sparingly since it can be a respiratory irritant. (It should be food-grade diatomaceous earth that's marked for livestock use.)

You can also add wood ash if you have some on hand. The wood ash should be free of chemicals, including lighter fluid. Make sure to remove any nails or hardware attached to the wood before burning it. The reason to add wood ash is that the charcoal in the wood ash contains vitamins and minerals, which can actually remove impurities from the body. Often when chickens are dust bathing in wood ash, they can be seen taking a few nibbles from the bigger pieces of burnt wood. Refresh the ingredients as needed to keep the bath effective.

If your dust box is outside of the coop and run, cats and other wildlife will find your bath attractive since it's like a big litter box. So it's a good idea to have a cover. Something easy like an old piece of hardware cloth cut to fit the space will keep critters out. Just remove it in the morning when your chickens are let out and replace it in the evening.

Bedding

Bedding is addressed in most of the beginner books, but most people still have questions about what makes the best bedding. The truth is that bedding is really a personal choice that has to be right for both you and your chickens. For you, it should be relatively inexpensive, easy to acquire where you live, and easy to work with as you move it from place to place. For your chickens, it needs to be absorbent as it catches and traps droppings. They'll also want it to cushion them, as birds jump off roosting bars and lay their eggs in nest boxes.

There are two bedding choices that are by far the most popular, pine shavings and straw, with regional and personal choices rounding out the pack.

PINE SHAVINGS

Pine shavings are the bedding of choice in my coop and meet the criteria of needs for both humans and birds alike. For humans, they are easy to get since they're sold at most major farm store chains and the mom and pop stores too. They're relatively inexpensive. They aren't very heavy. And they come packed in plastic wrapping so they store and stack well. For birds, they are absorbent but they don't hold the moisture in a wet mess; they dry out easily. In fact, if they're not overly dirty, you can leave the coop door open and let the air and wind dry out the chips. I've found a layer that's about 3 to 5 inches thick makes for an ultra-cushioning material for the floor and nest boxes.

There is debate in the chicken-keeping community about using cedar in the coop. Some say it can hurt a chicken's respiratory system. Others say this is not true. I think it makes sense to use caution here. I would not use cedar shavings in a brooder with baby chicks. But if your adult chicken coop is well ventilated and your chickens are free to roam, you could try mixing some cedar shavings with your pine shavings. That will lessen the impact of the aromatic oils and give you the insect- and pest-repelling properties of the cedar.

STRAW

Straw is also a popular choice. The downsides are that it can be harder to get since it's not readily sold at all stores, you need a proper vehicle to haul it, and straw can be harder to store without a barn or designated area in an outbuilding. But if you live in an area where you can easily get and haul straw, then it makes a great bedding choice. It absorbs well, but can be a little messy if it gets too dirty with droppings. It cushions well on the floor and in the nest boxes. With its hollow shaft, straw traps warmth and is a great insulator in the winter. It does have a bonus feature too: small kernels of grain are left in the bale, so your chickens will love picking through it for treats!

Be careful what you're buying because straw is not the same thing as hay. Straw is usually yellowish in color and consists of the dried stalks of grain plants after the grain heads have been removed. It has no nutritional value so it should not be used as a food source. Hay is usually greenish in color and consists of the grain heads that

Fresh bedding is a must for a clean coop and healthy birds. Pine shavings are a popular and widely available choice.

are removed from plants before they mature and die. Good hay does have nutritional value and is often fed to livestock.

Although these are the two main bedding options, there are alternative and local options. Some folks have great success with adding grass clippings, shredded newspapers, leaves, and sand. However, there are cautions and downsides to using these options.

> ## Note
> While some chicken keepers like shredded newspapers and office paper, I've found they don't absorb well—especially the glossy paper. The paper gets slippery, which can cause injury, and printed papers are full of ink, which can be toxic to chickens.

Grass clippings tend to hold moisture and start to smell. I have added grass clippings to my coop and found my chickens go out of their way to ignore them, and I end up having to remove them. If your chickens don't free range, grass clippings can be a nice treat and may be a little more enticing. Grass clippings tend to hold moisture and start to smell. I have added grass clippings to my coop and found my chickens go out of their way to ignore them. If your chickens don't free range, grass, clippings can be a nice treat and may be a little more enticing. But beware as they can cause an impaction and death. Always make sure your grass clippings come from yards that are free of pesticides and chemicals. Leaves are a treat in my chicken run or in a designated area of my yard. They do get messy, wet, and slippery, so they are quickly removed to the mulch pile once the chickens shred them and pick through for bugs and treats.

Sand is something I don't use as bedding in my brooders, coop, or attached run. Sand is expensive, hard to move, doesn't compost, and is messy. Think about the sand that families track inside after a day at the beach. That's not something you want to deal with at the coop! It's not cushioning either, especially in the winter during freezing temperatures. Likewise, it gets extremely hot to the touch if it's in an area without shade during hot summer days.

Cleaning the Coop

Chicken coops need daily cleaning, just like your home. Chicken coops need daily cleaning, just like your home. Think about ease of cleaning when you buy or build your coop. A well-designed coop should be easy to clean. In the long run, it will make for much happier owners and healthy chickens. This doesn't require a lot of work or time, but it's nice to refresh nest box bedding as it starts to show wear; maybe add some nesting box herbs as a nice treat. It's also a good idea to clean up the droppings area under the roost bars. That's the area of the coop bedding that gets the dirtiest. Removing and replacing that select dirty bedding saves you time and money because you don't have to clean the whole coop. The rest of the bedding doesn't get dirty as fast.

Just as you do at home, spring and fall are great times for a thorough coop cleaning. It's also a good time to make any repairs before more extreme weather sets in as the seasons change.

First take a hard honest look at your coop, making sure to check for areas that need some tending after the harsh weather. It's a good idea to head to the hardware store for all your fixing supplies once you know what jobs you need to tackle. Depending on your to-do list, you may want to set aside a day or whole weekend for repairs and for cleaning. Why not tackle both at the same time?

For a thorough cleaning, first remove all the old bedding. Then give your walls and nest boxes a good scrubbing. Leave all the doors and windows open afterwards to let the coop air dry. White vinegar and water work well for this task. Dish soap and water are fine to use too, but that combination requires rinsing, which can make a bigger mess. Caked-on chicken droppings can be tough to remove, so it's good to have a variety of tools on hand, including rubber gloves to protect your hands, a face mask so you don't inhale all the dust you'll kick up, and a stiff wired brush or a scraper for some serious poop removal. Also, make sure to get all those dusty cobwebs out of the corners and use some white vinegar and water to clean the window glass after you're all done. Once the coop is dry, make sure to add your clean bedding on the floor and in the nest boxes.

Tip

Many chicken keepers like to install a droppings board beneath the roost bars. This keeps the droppings out of the bedding and just needs to be cleaned daily. It's also a great way to check on the health of your chickens since you'll become intimately familiar with their droppings.

Biosecurity

Biosecurity is a word you hear often when you're around chickens, but what does it really mean? In a nutshell, it comes down to proper health and sanitation, so you don't spread diseases to or from your flock. Cleaning the coop is a big part of a practicing good chicken-keeping biosecurity. Here are a few other good practices to follow:

- Wash your hands before and after handling your chickens, their eggs, and their equipment.

- Keep wild bird feeders far away from the chicken coop.

- Quarantine all incoming birds for at least 30 days.

- Care for quarantined birds last to reduce their contact with your existing flock.

- Use designated coop shoes when you're in and around your coop.

- Do not let other chicken owners walk readily into your coops and chicken areas. Make sure they wash their hands and cover their shoes first.

Preparing for Emergencies

I usually have family members watch my home and animals when my family travels. As far back as I can remember, I've always prepared extensive sheets of instructions for them each time I left and

While a chicken emergency kit may seem excessive, when you need it you'll be thankful you have it!

then and called back to check on things. What I didn't prepare for are the times in a normal day that I couldn't get back home to take care of the chickens. I'm not talking about a major disaster like a tornado. But what about life's unexpected twists and turns, like if you're on a day trip with the family and your car breaks down. Or if you have to go to the hospital.

If this happened, most of my animals would be fine. They live in the house and have enough food and water for an extra day or so. But what about my chickens? They have to come in at night and go out in the morning. They need a little more care than my other animals. For years, I didn't have any instruction sheets for those times. Sure I

can call family members and let them know what to do. But it's easy to forget things when your routine is thrown off!

My solution was to take my vacation instructions and modify them to create chicken emergency care instructions. I printed them and covered them in contact paper for protection. These instructions allow anyone taking care of my chickens to have the vital information readily available to keep the flock healthy and happy. You never know what might happen. Even well-meaning family members can make costly mistakes. They may not remember exactly how many chickens you have. If the birds are free ranging and scattered, someone could come in and shut the coop

Emergency Chicken Care Instructions

List of Chickens
- Include names, quantities and brief description of each chicken.

  Ex. Thelma & Louise - (2 qty - Buff Brahmas - Gold with Black Necklace)

Feeding Instructions
- Include how much food to give and where food is given.
- List where food storage containers are located.
- Can include a list of favorite treats.

  Ex: Scratch grains are a favorite treat. They are located in garage in metal garbage pail. Please fill metal treat bowl with two scoops.

Emergency Contact Numbers

Alternative Caregivers

Veterinarian

Emergency Animal Hospital

Daily Routine
- Include time when coop opens in the morning. Include where food and water are located during free range time.

  Ex: Chickens are let out in the morning around 7:30 a.m., but this can be earlier or later. They are fed adult chicken food. It is stored in the garage in the metal garbage can.

- Include time when coop shuts at night and where to locate food and water during that time.

  Ex: Chickens free range all day and will put themselves to roost at night. This happens just before sundown. Shut and lock both the exterior coop and run doors after the chickens have roosted.

Chicken Emergency Kit
- Include where kit is located.
- Include quick list of supplies and possible uses.

  Ex: Latex Gloves to be used when caring for a wound.

  Neosporin should be applied after wound is cleansed.

Instructions hung in your coop and kept in your chicken emergency kit can help friends and family properly care for your chickens.

door for the evening thinking that everyone's inside and accidentally miss a chicken, leaving him or her without protection!

The following is a list of what I include in my instructions. I keep them in the chicken coop and I keep a copy indoors in my chicken emergency kit. This is just a guideline, and it can be personalized as needed. Remember to be brief but pertinent. While you give your chickens lots of love, care, and extra attention, this is just about the basics until you can get back to your flock.

List of Chickens: First give an overall count of how many chickens you have in total. Then break that list down if you have different breeds in your flock. You can list by type, making sure to give a general description of their overall appearance, the quantity of each type and their names if they're used to responding to them. For example: *Thelma & Louise* (2 qty.— Buff Brahmas—Large Birds, Gold with Black Necklace and Feathered Feet.)

Feeding Instructions: This is vital since your birds can't live without sustenance. Include where to get fresh water and how to give it. List where food storage containers are located along with how much food to provide and where the food should be placed. You can also include a list of favorite treats just in case your caregiver wants to go beyond the basics. For example: *Scratch grains are a favorite treat. They are located in garage in metal garbage pail. Please fill metal treat bowl with two scoops in the morning.*

Emergency Contact Numbers: If you have a veterinarian and a local emergency animal hospital, you'll want to include their phone numbers. Also include other alternate caregivers that can take over if needed. Sometimes even caregivers have emergencies and need relief too.

Daily Routine: It's important that your birds stay in as much of a routine as possible. Include the time to open your coop in the morning and what to do when the coop opens. For example: *Chickens are let out in the morning around 7:30 a.m., but this can be earlier or later. Their food and water can be placed on the patio in the shade. They are fed adult chicken food. It is stored in the garage in the metal garbage can. If you cannot check on them during the day, then do not let them free range. Open the door to the run and let them stay in the coop and run. Place the water along with the food in the run.*

Also include the night-time routine, so everyone gets to bed safely. For example: *Chickens free range all day and will put themselves to roost at night. This happens just before sundown. Count to make sure all chickens are present. Remove food from the run and store it in the garage. Shut and lock both the exterior coop and run doors after the chickens have roosted.*

Chicken Emergency Kit: Explain where the kit is located and give a quick list of supplies and possible uses if not self explanatory. For example: *Chicken emergency kit is located in the laundry room in the plastic box with orange handles. Latex gloves should be used when caring for a wound.*

The Coop Emergency Kit

We know emergencies happen without any warning. Just as we keep supplies on hand for human emergencies, it's good to have supplies on hand for chicken emergencies. This doesn't have to be elaborate, but it is good to keep all your chicken supplies outside of your family medicine cabinet. Even though some of the supplies may be the same, you don't want cross contamination. A nice plastic bin with a lid works well and should be stored in a cool and dry location so the active ingredients in the medicines don't degrade.

KEEP THINGS CLEAN

- Disposable gloves: These keep your hands clean and free of germs. Can be used when working with wounds, giving a bath or any other time you're in contact with bodily fluids.

- Saline solution: Can be used to wash out a wound.

- Cotton balls and cotton swabs: Used to apply topical ointment and clean out wounds.

- Paper towels

WASHING AND RINSING

- Hydrogen peroxide: Helps to clean a cut.

- Saline solution: Can be used to wash out a wound.

- Poultry care wound spray

Q&A: Coop Truth

How do I keep my chickens from making such a mess scratching and throwing food around?

People have different strategies to stop food messes and waste. Some use pellets instead of crumbles and say there is less mess. There are automated feeders that open only when the chickens step on a pad that pops the lid up. The food in the feeder section is covered by large mesh holes so the chickens can't get their feet up and scratch it out. The best rule of thumb is to adjust the height of your chicken's food and water so it's about the height of the back of the birds.

How do I get my birds to roost at night?

Roosting is an instinctual behavior to stay safe from predators. If you have an existing flock, they will show the newcomers where to roost by example. If you still have a bird or two that won't roost, first make sure there is nothing physically wrong with them. If they are not sick or hurt in any way, then go into the coop at dusk after the other birds have roosted and gently place the wayward chickens onto the roost bars. It may take a few tries, but they should get the idea.

Is it OK for my chickens to sleep in the nest boxes at night?

In the big scheme of things, it's fine if your chickens sleep in the nest boxes. It really won't hurt anything. But, most people answer this question with a resounding *no*. The reason for that is chickens poop throughout the night. So, a chicken sleeping in the nest box will make a mess. Unless you get up bright and early to change the litter, your other hens will lay their eggs in the soiled box and you'll have soiled eggs. If sleeping in the nest box happens occasionally, it's not the end of the world, but it's not something you want to encourage by letting it happen consistently. You can discourage your hens from sleeping in the nest boxes by gently removing them from the nest box each night and placing them back on the roost bar. They'll get the hint pretty quickly. You can also try closing your nest box curtains at night to hinder access.

Can I use bleach to clean my coop?

Bleach is a great sanitizer, and it's usually the first thing people grab when they're cleaning. You can use it, but you have to be careful because when it mixes with ammonia from the coop, the fumes can be toxic. A solution of 1-part bleach to 9-parts water dilutes the bleach enough while still keeping its cleaning power strong. But only use this when your birds are not in the coop and are not going to be in the coop for most of the day. Open all the doors and windows and make sure to let it air out properly as you're cleaning and as it dries.

COVER UP!

- Vaseline: Seals surfaces and protects against frostbite.

- Green Goo: An all-natural counterpart to Vaseline.

- Neosporin: Antibiotic ointment (do not use the pain relief type). Keep the spray and ointment on hand to deal with different types of wounds.

- Vet wrap and gauze pads: Can cover a wound to keep it clean and protected. Can use electrical tape or wound tape to keep covering in place.

TOOLS OF THE TRADE

- Scissors

- Towel: Can have many uses including wrapping and confining a chicken.

- Tweezers

- Cornstarch: Used to stanch bleeding. Styptic powder is also available in pet stores.

- Electrolyte powder for chickens: Can give a boost during hot weather or during times of stress.

Emergency kits can be expensive to put together and you may wonder if you'll ever need it at all. I can say from experience, my husband expressed the same concerns until the day we had a true emergency. I ran inside and grabbed the emergency kit and everything was right there at our disposal. Afterwards my husband came to me and told me what a great idea the emergency kit was and has even found things to add to it over time. It's one of those things that you don't need every day but when you need it, you'll be thankful you have it.

RESOURCES

Websites

COUNTRYSIDE NETWORK

www.countrysidenetwork.com

This is the digital home of four sister magazines: *Backyard Poultry, Countryside, Sheep!* and *Dairy Goat Journal*. Daily posts offer readers information on a variety of topics, including poultry, livestock, gardening, and more.

MY PET CHICKEN

www.mypetchicken.com

My go-to Internet resource when I first started raising chickens, My Pet Chicken has all kinds of beginner information along with a great selection of day-old chicks and hatching eggs, plus coops and chicken supplies.

THE INCREDIBLE EDIBLE EGG

www.incredibleegg.org

The consumer website for the American Egg Board, this site includes a great eggcyclopedia of chicken and egg terms. There are also lots of delicious recipes and nutrition facts to make sure you stay healthy.

NUTRENA CHICKEN & POULTRY FEED

www.nutrenaworld.com and scoopfromthecoop.nutrenaworld.com

There are two sites that I enjoy from Nutrena chicken and poultry feeds. The main site provides consumer guidance to help purchase the correct feed. This site also offers information for other pets and livestock. The Scoop from the Coop is Nutrena's less formal blog, with regular updates.

PURINA POULTRY

www.purinamills.com/chicken-feed

This is a support resource for Purina customers that helps them choose the feed(s) that are most appropriate for their flock. Purina experts offer advice and insights for correctly raising your flock.

FRESH EGGS DAILY

www.fresheggsdaily.com

This popular blog run by Lisa Steele is a great resource for information about raising chickens naturally. Topics include gardening, health, feeding, and all sorts of other information on raising a mixed flock of chickens and ducks.

TIMBER CREEK FARM

www.timbercreekfarmer.com

Created by Janet Garman, this blog explores farm life with various livestock including chickens, ducks, pigs, sheep, cows, and goats. It's a real-life account of how chickens fit into a farm.

POULTRY HUB

www.poultryhub.org/

Based in Australia, this website is a great resource for all things scientific when it comes

to chickens. There are wonderful graphics and explanations on everything from anatomy to proper poultry nutrition. There is also information about other poultry types which is good for the mixed flock owner.

CORNELL LAB OF ORNITHOLOGY

www.birds.cornell.edu

A wonderful resource for wild bird lovers. This website has an All About Birds online field guide and is home to Project Feeder Watch which is a citizen science program that monitors backyard bird populations.

USDA FOOD SAFETY INFORMATION—
Shell Eggs from Farm to Table

www.fsis.usda.gov/wps/portal/fsis/
topics/food-safety-education/get-answers/
food-safety-fact-sheets/egg-products-preparation/
shell-eggs-from-farm-to-table/

This is an incredibly detailed portion of the United States Department of Agriculture website that contains everything from the history of the egg to explaining proper egg handling procedures to a handy egg storage chart.

UNIVERSITY OF KENTUCKY
College of Agriculture, Food and
Environment—Poultry Extension—
Small and Backyard Flocks
www2.ca.uky.edu/smallflocks/
An easy-to-understand website that covers all
aspects of chicken keeping including nutrition,
health, and breeding.

EXTENSION—For Extension Professionals
and the Public They Serve
articles.extension.org/poultry
This website serves the cooperative extension
system and has a wonderful section dedicated to
small and backyard flock owners.

UNIVERSITY OF ILLINOIS
EXTENSION—Incubation and Embryology
extension.illinois.edu/eggs/
This is an online resource for elementary
through high school teachers that explores the
importance of the egg and classroom hatching
projects. The resources section is great for backyard chicken owners.

Books and Magazines

- *Backyard Poultry* magazine by
  Countryside Network
- *Hobby Farms* magazine by Lumina Media
- *Chickens* magazine by Lumina Media
- Brock, Todd, Dave Zook and Rob Ludlow.
  Building Chicken Coops for Dummies:
  Hoboken, NJ: Wiley: For Dummies. 2010.
- Damerow, Gail. *The Chicken Health Handbook.*
  North Adams, MA: Garden Way Publishing /
  Storey Publishing. 1994
- Damerow, Gail. *Storey's Guide to Raising
  Chickens.* North Adams, MA: Storey
  Publishing, 2010

Editors

- Gauthier, Julie and Rob Ludlow. *Chicken
  Health for Dummies.* Hoboken, NJ: Wiley:
  For Dummies, 2013.
- Steele, Lisa. *Fresh Eggs Daily.* Pittsburgh, PA:
  St. Lynn's Press, 2013.
- Smith, Miranda. *Your Backyard Herb Garden.*
  Emmaus, PA: Rodale Press, 1997.
- Editors at Reader's Digest. *The Complete
  Illustrated Book of Herbs.* New York, New York:
  Reader's Digest, 2016.

ACKNOWLEDGMENTS

When I think of my chicken-keeping journey, it's not just about me. It's about my husband and my daughters too. Although I tell the story, it's really a story about all of us. We've been together as a team on this chicken-keeping path from the beginning. From the joy of brooding new chicks, to the countless hours and trips to the hardware store to buy supplies, to daily egg collecting, to the heartbreak of losing flock members, we've made memories over the years that will last a lifetime. I'll never forget things like "Curse you, Hoppy!" or Henrie sleeping upside-down in little girl arms. From the bottom of my heart, thank you to my husband and daughters for sticking with me and supporting me, especially as I wrote this book. You gave me the time to write and never complained if I was distracted or not spending enough time with you. Thank you. Thank you. Thank you. I love you guys!

To my colleagues at *Backyard Poultry* magazine and *Countryside* magazine, I enjoy working with you every day. It never gets old bringing the joy of poultry keeping to the world. We have such a good time and I look forward to many more creative endeavors.

To Lisa Steele, thank you for introducing me to the folks at Voyageur Press. Your willingness to help make someone else's dream a reality warms my heart.

To Thom O'Hearn, I owe you a huge thank you. As an editor myself, I don't often have the privilege of having someone else review my work and help me grow as a writer. I appreciated the frank advice and the second set of eyes more than you know.

To Chris Cone, my wonderful book photographer. Thank you for spending countless hours in my backyard bringing the beauty of my birds to print. Your pictures truly capture the personalities behind my feathered friends. I'm sure my photo shoots were some of the most unusual you've ever had, and certainly the most fun!

To Dr. Sherrill Davison, thank you so much for all the time you put into reading and reviewing this book. It's a privilege to work with someone that is so dedicated to the poultry community and the health of backyard flocks.

To everyone that reads this book, I hope you enjoy your flock as much as I enjoy mine. I hope this book adds to your chicken-keeping repertoire and helps your flock to stay happy and healthy.

INDEX

ABOUT THE AUTHOR

Pam Freeman is the editor of *Backyard Poultry* magazine and *Countryside* magazine.

After she received four Silver Laced Wyandotte chicks from the Easter Bunny, her flock quickly grew and Pam launched pamsbackyardchickens. com. In the years that followed, she hand-raised chicks, nursed chicks and chickens back to health, and experienced the entire lifecycle many times over. During that time, Pam also joined the Countryside Network, where she is now an editor managing a roster of her fellow chicken-keeping writers. Pam is a resident "Ask the Expert" columnist for *Backyard Poultry* magazine and continues to write regular posts about chicken keeping and homesteading for *Backyard Poultry* and other publications.